Macro-engineering and the Earth:
World Projects for the Year 2000 and Beyond

A festchrift in honour of Frank Davidson

It is a massy wheel
Fixed on the summit of the highest mount,
To whose huge spokes ten thousand lesser things
are mortis'd and adjoin'd; which when it falls,
Each small annexement, petty consequence,
Attends the boist'rous ruin.
Shakespeare: *Hamlet* II, iii

First published in 1998 by
HORWOOD PUBLISHING LIMITED
International Publishers
Coll House, Westergate, Chichester, West Sussex, PO20 6QL
England

British Library Cataloguing in Publication Data
A catalogue record of this book is available from the British Library

ISBN 1-898563-59-4

Printed in Great Britain by Martins Printing Group, Bodmin, Cornwall

Macro-engineering and the Earth:
World Projects for the Year 2000 and Beyond

A festchrift in honour of Frank Davidson

Uwe Kitzinger, CBE
President
International Association of Macro-Engineering Societies
and
First President of Templeton College
Oxford

and

Ernst G. Frankel
Professor of Ocean Engineering
Massachusetts Institute of Technology
Cambridge, Mass
USA

Horwood Publishing
Chichester

Editorial Note

As is so often the case, the names on the cover are those of two men. As is just as frequent, behind the two male names there hide the names of two other essential contributors to the joint task.

Ernst Frankel's office assistant, Sheila McNary, has been most helpful to us both and generous with her time well beyond the call of her MIT duties.

Cherie Potts of WordWorks has, over the past seven years, become in effect Editor Extraordinary to the field of Macro-Engineering. Without her unfailing patience and good humour, let alone her eagle eye for accuracy, this volume could not have been produced in so short a time.

The editors also wish to thank for their diverse help and support the Board members of the International Association of Macro-Engineering Societies and of The American Society for Macro-Engineering.

Needless to say the views expressed in the different chapters are the personal views of their authors only, and commit neither the editors nor the bodies with which the authors are associated.

Table of Contents

Tribute to Frank Davidson

Rarely does an individual become identified, to the exclusion of others, as the source of an activity or area of study. Even more rarely is that exclusion reasonable or justified. However, that has happened with what we call macro-engineering, defined by Davidson and Salkeld as "nothing by the study, by the best minds in all relevant disciplines, of the largest engineering enterprises which society can accomplish at any given period of time."[2] It has become literally the personal property of Frank P. Davidson.

Starting with the first three macro-engineering symposia in 1978, 1979, and 1980 co-sponsored by AAAS and AIAA, there has been a significant amount of interest and discussion about the issues raised by students in this field. We have had a number of conferences, many books, particularly Frank's *Macro*,[3] innumerable papers and the formation of The American Society of Macro-Engineering (TASME) and the International Association of Macro-Engineering Societies (IAMES), both co-founded by Frank. Generally, those involved in the field believe that at times very large-scale projects are not only appropriate but frequently represent the optimal solution to many problems and should not be avoided just because they may be particularly difficult and challenging. Further, large projects deserve special study since they represent an order of difficulty vastly different from smaller projects. The emphasis has been on a multi-disciplinary approach with special concern for their social and environmental impacts.

One might wonder how a Harvard Law School graduate became the "father" of a new field of study in engineering. It is really not so surprising to those of us who know Frank. The requirements for this role are both unique and numerous, and by background, training, and personality he is an ideal macro-engineer.

The first requirement is that one must be an optimist about the ability of man to solve problems and control his destiny, without falling into the sin of *hubris*. Not always an easy trick. I believe that Frank is convinced that if the proper group of experts is brought together and they truly listen to one another, great things can be accomplished. The important thing is to get on with it and quit muddling about looking for excuses for inaction. The key is to be sure that all of the requisite skills, engineers, lawyers, investment and commercial bankers, public and government relations experts,

---

[2]   Salkeld, R., Davidson, F.P., and Meador, C.L. (eds.), *Macro-Engineering: The Rich Potential*. New York: American Institute for Aeronautics and Astronautics, 1980.

[3]   Davidson, F.P. *Macro*. New York: William Morrow and Company, 1983.

environmentalists, sociologists as well as skilled managers, are utilized in the planning process. While there are many problems in the implementation of a macro-project, engineering and technological problems are rarely the cause of failure. Public opposition, governmental regulations, financial shortfalls, and environmental problems are the more likely culprits.

Second, there must be an awareness of the complexity of large-scale systems. The degree of difficulty in implementing a very large-scale project versus the problems involved in small projects is disproportionally greater with the risks of failure far more likely. Frank was for many years Chairman of the System Dynamics Group at M.I.T., a group concentrating on the behavior of very complex systems. He may also have been the only person to teach a course in failure at the college level.

Successful communication is another key element in implementing a large-scale project. Communication not only between promoters of a project and the public, but also communication among the various disciplines involved in implementation. It may very well be the most critical area for success. Frank has recognized this and made a point of emphasizing its importance in his books and papers. This could very well be because no one could be a better communicator than Frank. Not only does he seem to know just about everyone who has any interest in large projects, but he has shown the ability to persuade them to pool their knowledge and resources with others in TASME and IAMES.

Frank has been a seductive Pied Piper for students and younger practitioners of this field. He has encouraged many of this group to participate actively in various conferences and in TASME and IAMES. Considering how many meals he has provided, not only for the young but also for the older crowd, it is not surprising that everyone responds promptly to his calls for participation in macro-engineering activities.

In Frank Davidson, macro-engineering has found the ideal advocate. His qualifications as strategist, publicist, and promoter could not be better. This volume is designed to spur him to even greater efforts and most definitely prohibit him from resting on his laurels simply because he is now a mature octogenarian.

Wallace O. Sellers
April, 1998

Tribute to Frank Davidson

Frank Davidson is one of those remarkable individuals whom we meet only rarely in the course of a lifetime.

Yesterday evening here in Paris, my wife and I were guests of Frank Davidson and his lovely and engaging French wife, Izaline. Other of Frank's friends at the dinner included a prominent French sculptor and his author/artist wife, a former French government official who had graduated at the top of his class from the famed Ecole Polytechnique, a retired professor and his poetess wife, and a renowned international banker.

In the course of the evening, the conversation covered an immense scope of subjects, all touching in some way on Frank's wide-ranging global, professional, and personal interests. The multifaceted discussions touched on such diverse areas as sculpturing, poetry, art, international financing of major projects, a proposed bicycle route from Paris to Moscow, and the early days of the Channel Tunnel project as originally conceptualized by Frank. Almost without a pause, we then considered current global environmental, transportation, telecommunica-tions and energy issues and the formation of an international institute to study these, turning next to modern designs for electric bicycles and finally to books recently written by various of the dinner guests.

This was but a typical dinner for which Frank is so renowned, each in its own way exemplifying the far-ranging interests and societal contributions of this beloved friend to so many in such diverse professions and occupations around the world.

I have had the enormous privilege and pleasure of knowing Frank now for almost thirty years. He is, simply put, one of a kind. Not only has he helped conceptualize a multitude of macro-engineering projects, such as the Channel Tunnel, vast energy systems and major high-speed transportation projects, but he has lent his continuous, dynamic support to a multitude of scientists, engineers, and entrepreneurs on the cutting edge of evolving macro-concepts.

As a leading writer and sponsor of meetings throughout the world, Frank has, over the course of a long career, provided intense, on-going personal leadership and inspiration in conceptualizing imaginative programs for proposed macro-projects. He, as well, has been a combat soldier, an international attorney, a reflective academic, an avid horseman, a mountaineer, a businessman and promoter developing the early work on the Channel Tunnel and other important projects, as well as a prolific author.

For some years Frank was associated with the renowned Dr. Jay Forrester's program at MIT involving System Dynamics. One day during one of the important functions hosted by Frank at MIT, I overheard Dr. Forrester remark, "Frank really adds something very special around here." That succinct remark captured in a nutshell Frank's unique attributes.

Frank Davidson does indeed add something very special to whatever and whomever he touches and that particularly includes the exciting field of big global projects, which so dramatically affect the course of much of human progress throughout the world. His many friends are indeed very privileged to continue to be so profoundly impacted and inspired by this enormously interesting and talented individual.

Cordell W. Hull
April, 1998

Preface - 1

This planet has major problems.

We face sharp further growth in population and continuing gross inequalities in health and health care, education and employment opportunities, in land, food, and other resources. We face a deteriorating natural environment from the seas and the rainforests to the ozone layer, a shortage of fresh water, and the undoubted need now to harness vast sources of clean inexhaustible energy to replace and supplement traditional sources.

Major problems cannot be met without major solutions.

Some solutions may involve the use of millions of "small and beautiful" devices harnessed together, like the helical turbines invented by Alexander Gorlov to dispense with dams and use slow flows of water. Others need world-wide agreements -- as on the harvesting of solar energy from space championed by Peter Glaser. Seamless global transport linking shipping and aircraft, tunnels and magnetically levitated superspeed trains make sense in the long run only on continental and intercontinental scales.

So major solutions today virtually necessitate internationally concerted action. Nothing less can make the world safer from intentional or accidental nuclear, chemical, and biological dangers. Nothing less can ensure a more rational and equitable use of water supplies through schemes like the fresh water conduit from the Rhone to the African desert examined by Ernst Frankel. Nothing less can attempt a credible system of environmental monitoring by satellites or a strategic defense against asteroid impacts on the earth -- as suggested by John Landis -- and against other natural or human-made large-scale disasters.

If the search for solutions to such macro-problems is termed "macro-engineering", the label "engineering" should not blind us to the essential character of such activity. Where macro-projects involve still unresolved problems of pure engineering, they are perhaps best not attempted at all. For such major under-takings face far more serious issues than finding technical solutions to unresolved mechanical or chemical or electronic problems. They face the challenges of macro-finance, macro-management, macro-diplomacy, and the management of the inevitable complex array of externalities -- the indirect by-products and often far-reaching unintended secondary effects in all sorts of domains.

This cats-cradle of interacting problems forces on us an inter-disciplinary and holistic approach. And what is more, it usually forces on us tough ethical dilemmas -- moral choices between contrasting values, between different groups, between the earth and its inhabitants, between the present and the future.

That is why we consider major projects in whatever sector -- be they below the seas or above the skies, in mining minerals or in satellite communications -- to be qualitatively different from minor projects in their own specific sector, and why we treat them as generically related much more to each other. That is why we believe it to be crucial for the twenty-first century to transfer experience (whether from failure or success) between technically quite disparate macro-projects, and to build macro-management theory and formulate "best practice" on the basis of that experience.

Macro-projects are achieved by coalitions coming together and staying together to achieve a specific functional purpose. Their common prerequisite is a sustained political will. And then it is in the management of the multiple interfaces between the participating bodies -- reflecting diverse disciplines, often clashing organiza-tional cultures, and sometimes acutely conflicting interests -- that we run the greatest risks of failure.

The management of macro-projects is thus above all an exercise in political and organizational leadership and in the management of complexity in the relations sometimes between individuals, usually between very disparately organized bodies of people.

We need a new focus on bold and generous initiatives, within fractured societies (like the civilian conservation corps advocated by Janet Caristo-Verrill), between nations in the same region, and indeed between the peoples of the planet for the sake of its future as a human habitat. In such re-focusing of human endeavor we see one long-term hope for a more collaborative and peaceful world.

That is why this is a subject not merely for the lawyers and financiers, who have to frame complex international contracts assigning responsibilities and risks, but even more for the political and social scientists, the students of human nature and of the impact of organizational structures on human groups.

The essays selected here reflect this approach. They follow on from the Brunel Lecture series published earlier.[1] They look back over some of the achievements of the twentieth century, and forward to the challenges -- and to the promise -- of the twenty-first.

To face the planet's unprecedented problems will require realism, vision, and courage -- realism in the assessment of the costs of inaction as well as the

---

[1]   Davidson, F.P., Frankel, E.G., and Meador, C.L. (eds.), *Macro-Engineering: MIT Brunel Lectures on Global Infrastructure*. Chichester: Horwood Publishing, Ltd., 1997.

Davidson, F.P. & Meador, C.L. (eds.), *Macro-Engineering: Global Infrastructure Solutions*. Chichester: Ellis Horwood Ltd, 1992.

costs of failure, vision to imagine the at present almost unthinkable, and courage to risk ridicule today and intense effort tomorrow.

This volume of essays is dedicated with great affection on his eightieth birthday to Frank Davidson, who has never lacked realism, vision, or courage -- and has had a seminal role in framing the overarching concepts and stimulating the international studies that have already, over the past two decades, made a difference to humankind's recent record in the achievement of major projects.

Uwe Kitzinger
President, International Association
of Macro-Engineering Societies

Preface - 2

With this volume we pay our respects not only to one of the greatest technological visionaries, one who uses his compassion and concern for the advancement of humanity, but one who also encourages people everywhere to think the unthinkable and imagine the benefits of achieving the assumed impossible. He has a knack for bringing together people of diverse backgrounds and interests, and fostering multi-disciplinary considerations in solving large-scale problems. He always feels that he can get things done that way -- change the world as we know it. It is people like him who really do. They are not afraid to suggest unpopular, unconventional, and presumably unfeasible projects, because they know that only by envisioning a better future can we move toward it -- no matter what the temporary constraint.

In this volume we offer a small sample of ideas, comments, proposals, reviews, and projections of large-scale macro-engineering projects which represent the state-of-the-art at the end of this century, with brief summaries of achievements during this century and suggestions for the next century.

Ever more rapidly advancing technology, the need for solving the gap in human conditions, and the new globalization of the world now require us to consider macro-engineering solutions which were thought to be economically unattractive, politically incorrect, and technologically risky, with due concern for the improvement of the environment of human living standards everywhere.

Large can not only be beautiful but is now necessary to achieve a better life and a secure future for all. We will have to merge projects into macro-projects to achieve the benefits in technology and economy that global merging of companies now attains. This is the way to the future and to the solution of the challenges of the next century in which people will have to advance while conserving resources for mankind's future.

Macro-engineering projects could simply be defined as very large, very expensive, very complex, or very advanced projects. But to be a true macro-engineering project, it must solve macro-problems. In other words, it must make really major contributions toward advancing civilization, standards of living, alleviation of poverty, and knowledge as well as economic growth. It must add significantly and permanently to the good of mankind and the world we inhabit.

In macro-engineering we are not only or even mainly concerned with large-scale and breakthrough technology or the size of the project in technical, financial, or dimensional terms, but with the long-term impact of the project. It must have a lasting effect and cause permanent change.

Macro-engineering projects fall into many categories and perform diverse functions. Increasing globalization in infrastructure, manufacturing, and services has caused project scales to increase substantially, a trend expected to continue and in fact accelerate. The next century should encourage a completely new type of macro-project -- the world project -- in which the international community-at-large joins to solve worldwide problems, from the greenhouse effect, global telecommunications, and supersonic or superspeed transport, to global energy transmission networks, regional water supply systems, global food banks and distribution systems. We really are at the threshold of a macro-engineering world where big will be good and beautiful.

Ernst G. Frankel
Professor of Ocean Engineering
MIT, USA

Introduction

The Twentieth Century Revisited

Ernst G. Frankel

Professor of Ocean Engineering, M.I.T.

This century was the most successful ever for mankind, notwithstanding two world wars, huge political upheavals, the dismantling of colonialism, the failure of communist and other socio-economic experiments, and the quintupling of the world's population. Absolute poverty has decreased from 70% to 30% of the world's population, life expectancy has more than doubled on average, living standards have vastly increased, and both public safety and health have greatly improved.

We started this century with a divided world. Enslaved colonial countries, a divided Europe, an Asia highly isolated or controlled by outside powers. Africa had been divided among the European powers, and even South and Central America, though independent, were largely ruled by despots. The United States was largely an agrarian country which had not yet exploded onto the world scene as a major player.

Canada and Australia were part of the British Empire and were used to provide raw materials and food from their vast territories. The British ruled the seas and thereby international trade, which amounted to a bare 8% of the world's gross product. There were few public services and technology was only beginning to emerge from the shadow of the Industrial Revolution in the 19th century - an era driven by coal-fired steam power.

Electricity was only sparsely used; Marconi sent his first transatlantic radio message in 1901; and Orville and Wilbur Wright successfully flew the first powered aircraft in 1904. Gasoline and diesel engines started to be used in automobiles, yet locomotives continued to be steam-driven. Per capita electric power consumption was less than one percent of today's consumption. Most electricity was generated by coal-fired steam power plants. Wired telephones became common in such urban areas, but urban private transportation continued to be a mix of horse-driven carriages and gasoline-driven automobiles. Public transportation was provided by electric trains and gasoline engine buses.

Many ships were driven by coal-fired steam engines though sail power remained an important source of propulsion until about 1910. Petroleum was not really of any importance, providing a paltry 3% of fuel. Inter-urban transport was largely by rail and every large coastal city had a port which served only its immediate surroundings. Homes were heated with coal or wood, and air conditioning was not yet in use except for a few experimental applications.

Even though some manufacturing was concentrated in small areas that had comparative advantages, such as cheap hydro power, access to raw materials, or abundant labor, most manufacturing was on a relatively small scale that just met meet local needs. The same applied to most agricultural production, which provided mainly locally grown food to the local markets. Few agricultural goods were transported over large distances, except for special produce only producible in certain climatic zones such as tropical fruits, coffee, tea, jute, and so on. As a result few agricultural products were processed.

Most economic activities were on a small local scale. Ships were individually owned and operated and few, except for passenger ships, provided regular fleet service. Most transport investment and operations were private as were most other supplies. Early in this century the need for public distributed networks of electric power, telephone, water, gas, sewer, and similar conduits brought local or regional government into play which assumed the investment and distribution or operation of the supply systems, and later quite often also built or acquired the basic suppliers: the power plants, telephone switchboards, water pumping, and so forth.

It was only in 1915 that Alexander Graham Bell transmitted the first telephone message from New York to Dr. Thomas Watson in San Francisco. Yet the beginning of this century was an age of great discovery. In 1915 Einstein espoused his General Theory of Relativity, and in 1909 the plastic age began with the discovery of Bakelite.

Just a few years earlier, work had begun on the Panama Canal and the first railroad tunnel was built under the Hudson River. Amundsen reached the South Pole in 1911 and Edwin P. Hubble discovered a distance indicating variable stars in the Andromeda nebula, a major astronomical discovery. Insulin was discovered and insecticides were used for the first time in 1924.

Television was first transmitted in 1925, but it took another twelve years before it become commercially useful. Flight became commercially attractive over comparatively short distances by airplane and longer distances by lighter-than-air craft or Zeppelins. In 1927 Charles A. Lindburgh flew his monoplane non-stop from New York to Paris in 33.5 hours and the 15 millionth Model T Ford had been produced. Ford had revolutionized the manufacturing industry by developing in-line mass production of a standard product, the Model T. His example was soon followed not only by other carmakers but also other manufacturers. This resulted in a drastic reduction in the cost of manufactured goods, but it also radically changed the workplace and the work environment.

In 1931 the first long distance submarine, the *Nautilus*, navigated under the Arctic Ocean. This was the age when huge new bridges, tunnels, and dams were built. It was also the time of intense research into the nature of fundamental atomic particles. By this time oil had become an important fuel and the first long-distance oil pipelines were built in the U.S. and in the Middle East. Refineries began to spring up in the major oil-consuming countries of Europe and America, and oil provided over 25% of the fuel used by society.

In 1936 the Hoover Dam on the Colorado River in Nevada, the world's largest dam creating Lake Mead, the world's largest reservoir, was completed. The Zeppelin was competing with ocean liners for transatlantic passenger traffic. Passenger liners grew to over 50,000 tons displacement and 800 feet in length, with a capacity of over 2,000 passengers. In freight transport, the first specialized liquid bulk tankers and dry bulk coal carriers entered service. In 1937 Frank Whittle built the first jet engine - to some ridicule - and the engine only entered service at the end of World War II, too late to make a difference.

It was just before the outbreak of World War II that Joliet-Curie demonstrated the possibility of generating huge amounts of energy by splitting the atom. The same year Igor Sikorsky built the first helicopter. In 1941 The Manhattan Project, a true macro-engineering project of intense atomic research, was started. In 1942, Henry J. Kaiser developed techniques for the mass production in just 30 days of 10,000 ton ships called Liberty Ships, a process which usually required 12-18 months.

At the same time, penicillin was first successfully used in the treatment of chronic diseases. This was also the era for use of large hospitals and health care complexes. The year 1945 witnessed the dawn of the atomic age with the detonation of the first atomic bomb on July 16 in New Mexico. The first pilotless rocket missile was flown in 1946, the same year a weapons-grade atomic bomb was detonated on the Bikini Atoll.

Supersonic speed in flight was first achieved in 1947, the year the transistor was invented, which really opened the electronics age. In 1948 antibiotics were first developed for effective use. Although radar was a secret weapon developed by the Allies during World War II, its first commercial application was in Liverpool in 1948. The first transcontinental jet flights were in 1949, the year the first guided missile was launched and flew 250 miles high. In 1954 the first electric power was produced from atomic energy in Arcon, Idaho.

In 1956 transatlantic cable telephone service was first inaugurated. Sputnik (3,000 pounds) was launched in 1958, the year the U.S. nuclear submarine *Nautilus* passed under the ice cap at the North Pole. In 1959 the first U.S. nuclear-powered merchant vessel, the *USNS Savanna*, was launched.

While the first sixty years of this century were a time of discovery and scientific breakthroughs, the last forty years has found man increasingly interested in innovation and technology development or their applications. Although scientific discovery and the search for knowledge continued, more emphasis was given to solutions to problems. This was largely due to increasing demands by a rapidly growing population in newly independent developing countries for access to technology, combined with the demand of people in developed countries for improvements in their quality of life. The world's population had more than doubled since the beginning of the century and would more than double again before its end.

There was also increasing concern about the need to alleviate hunger and poverty as well as find some way to reduce the vast disparities in income and living standards between rich and poor and between developed and developing nations. Demand for services and consumables exploded. The capacity of transport, energy, water supply, communications, health care, and educational systems had to grow not only at the rate of growth of the population but at many more times to fill the gap left by lacking or inadequate service capacity. At the same time, consumption of food, apparel, appliances, and other personal investments, such as housing, skyrocketed as people demanded access to such things worldwide.

To cope with increases in demands for energy, transport, water supply, and other services that were exploding at a rate exceeding 10% per year required huge infrastructure investments. Although there were many true macro-engineering projects before or during the first sixty years of this century, worldwide demand for major projects started to skyrocket in the last forty years of this century. This period is therefore truly the dawn of the macro-engineering age.

Huge hydroelectric, nuclear, and fossil-fueled power projects were started, as were large water supply and drainage projects. The Guri, Igazu, and Aswan dams are examples of these developments, as is the largest of all such projects - the Three Gorges Dam now under construction in China. Large submerged aqueducts to transport water from northern to southern California are other examples, and the new manmade river projects in Upper Egypt and in Libya are further illustrations.

To reach new oil reserves, huge offshore drilling platforms, some more than 380 meters tall, have been constructed. Transcontinental and transocean pipelines as much as six feet in diameter are transporting gas and liquid petroleum over thousands of kilometers in the U.S., Asia, and Europe.

Macro projects in transportation include fixed or floatable artificial island ports and airports, high- speed train networks in Japan and France, intelligent highway systems, and continental traffic control systems. Transport infrastructure has advanced to take advantage of major breakthroughs in vehicle as well as operating control and communications technology. Pilotless aircraft and unmanned ships are technically feasible today. Supersonic aircraft with capacities of up to 500 passengers are in use, and tankers with cargo-carrying capacities exceeding half a million tons serve major trade routes. We use containerships that can carry over 6800 TEU containers at speeds of 25 knots, and our new cruise liners are veritable floating cities. One plan envisions building a 250,000 ton displacement World City vessel capable of accommodating 6,800 passengers.

These developments are now driving the construction of super ports and super airports which in turn become efficient intermodal hubs. To achieve this, modal transport companies are merging into ever-larger modal and intermodal enterprises which often involve global combines and alliances. As a result, transportation is truly becoming interactive and globalized. Many of these developments were made possible by the revolution in computing and information technology. Global scheduling, routing, positioning, tracking, booking, and control are now possible in real time which permits effective operational management of operations and vastly reduces routine management functions. Intermodal integrated transport of passengers and freight has truly assumed macro-engineering proportions, a trend that is expected to continue well into the next century.

The most important advancements have probably occurred in computing and telecommunications. Here technology emerged from simple information transfer and recording into veritable worldwide communications and control systems that permit instant real-time paperless transactions of all sorts. Advances in computers and telecommunications have radically changed the way we live, do business, manufacture, transport, transact, and most importantly, communicate. The Port of Singapore, as an example, has nearly 100,000 transactions per day involving transport, transfer, clearance, and more - all done without paper.

We control the position and traffic and perform millions of transactions per hour electronically. We developed space craft, landed on the moon in 1969, and now have reusable shuttles which allow us to transport people and material into space, and foreshadow future space travel between earth locations in minutes instead of hours or days.

Many factories are now manned by robots and agriculture is planned and controlled by satellite signals. We built the English Channel Tunnel and a space station. Other long subsea tunnels were built in Japan. Irrigation and flood control systems in China, Europe, and the U.S. have become huge macro-projects. We are exploring the deepest oceans and have developed submerged production platforms to extract oil, gas, and (in the future) minerals from the ocean bottoms. All of these were and still are driven by the macro-engineering advantage - the opportunity that size and large-scale cooperation provides. The merging of talents and capabilities, as well as the globalization of organizations in financing, engineering, and resource marshalling not only made these macro-projects possible but fostered a new sense of cooperation and a willingness to share risks, costs, and benefits for the greater good of mankind.

The last forty years have wrought many changes in almost all human activities. Most of these changes were technology-driven as we learned to effectively apply scientific discoveries. The most important characteristics of this period were the increasingly fast rate of technology development and application. In fact, few technologies for the length of their economic lifespan. Most are quickly overtaken by more advanced or superior technology.

The fears of forty years ago - that mankind would run out of food, fuel, and other resources needed to meet growing demands - have been dispelled as we find ourselves in an increasing environment of plenty with steadily declining raw material prices and abundant petroleum, food, and minerals. Yet, we have not learned both effective use or distribution of resources and their cost, availability, and use are therefore diverse, and sometimes scarce in some parts of the world.

At the same time we are finally becoming aware of our impact on the physical environment. Major projects have been undertaken to reduce the greenhouse effect, ocean pollution, and other impacts, and large resources are being used to reduce man's impact on his own environment.

Most importantly, we have learned that we must deal with these and other problems and opportunities on a regional, often multinational, or even global scale because impacts and opportunities know no borders. This has resulted in large increases in the scale of infrastructure, finance, manufacturing, agricultural, environmental protection, and other projects. The projects that could truly be termed macro-engineering projects now double nearly every ten years in both number and size.

The last forty years can therefore be said to have become the period in which macro-engineering was not isolated to a few projects, but where it became the preferred approach for capturing not just economies of scale but also the true operational, technological, environmental, and social benefits that would have been lost in a micro-approach to the same problem.

Section One

ENERGY

1

Turbines With a Twist

Alexander M. Gorlov
Professor of Engineering, Northeastern University

INTRODUCTION

This chapter describes the helical turbine as an efficient new instrument for converting the kinetic energy of hydro streams into electric or other mechanical energy. A multi-megawatt project is proposed, conceived as an ocean power farm equipped with a number of helical turbines, along with a floating factory for *in situ* production of hydrogen fuel by means of electrolyzing ocean waters. Besides mega hydro-power farms, mini-power stations with helical turbines of a few kilowatts each are also proposed as possibilities for small communities or even individual households located near tidal shorelines or river banks with strong water currents. No construction of hydro dams is necessary for such applications. As in hydro power plants, compact helical turbines can be used in wind farms instead of huge, conventional propeller-type machines. The advantages of such a design for future wind power systems are described in this chapter. The following illustration summarizes prospects for application of the helical turbine in various energy systems.

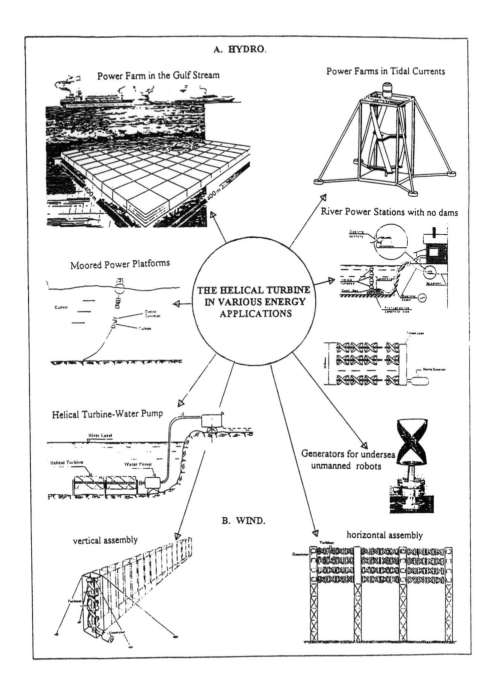

A. HYDRO.

Power Farm in the Gulf Stream

Power Farms in Tidal Currents

River Power Stations with no dams

Moored Power Platforms

THE HELICAL TURBINE IN VARIOUS ENERGY APPLICATIONS

Helical Turbine-Water Pump

Generators for undersea unmanned robots

B. WIND.

vertical assembly

horizontal assembly

THE POWER OF OCEAN STREAMS AND OTHER ULTRA LOW-HEAD HYDRO SOURCES

The kinetic energy of ocean streams (such as the Gulf Stream or the Kuroshiwo current near Japan), as well as tidal and monsoon streams, is tremendous. However, the absence of an efficient, low-cost and environmentally friendly hydraulic energy converter suited to free-flow water is still the major barrier to exploiting this renewable energy source. Another well-known barrier to the development of renewable energy is, unfortunately, the low cost of oil which remains the principal component of world energy production. But it is time to realize that world reserves of oil are limited and rapidly dwindling. Moreover, since hydrocarbons such as oil and coal are of considerable importance as raw materials for industry, especially for future generations, their burning should be limited. And the concept that "life is hard but it's fortunately short" will not help much here.

For decades scientists and engineers have tried unsuccessfully to utilize conventional turbines for low-head hydro. The efficient hydraulic turbines in high heads become so expensive in applications for low- and ultra-low-head hydro-electric stations that only the most modest development of this kind can be found in practice.

Three principal types of hydraulic turbines are presently used for harnessing hydropower, namely: Kaplan, Francis, and Pelton and some of their modifications, such as Bulb or Straflo turbines. However, as can be seen from Exhibit 1.1, the

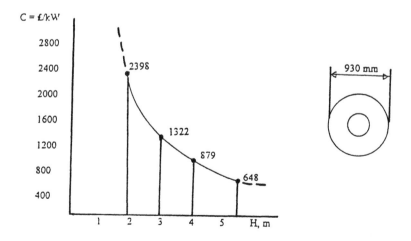

Ex. 1.1 Unit cost of Kaplan turbine vs. hydraulic head (by British manufacturers)

low-head hydro is the same as using a racing car instead of a tractor for picking up crops although they can both develop the same power.

The energy of fluid flow is described by Bernoulli's equation:

$$z + \frac{p}{\rho} + \frac{V^2}{2g} = const$$

where

(p/ρ) = the energy part that is caused by external pressure (water head),
$V^2/2g$ = the kinetic energy component, and
z = the fluid elevation with respect to the reference axis.

When z is taken as an origin of coordinates $z = 0$.

Conventional turbines (except the Pelton turbine) are designed to utilize mostly the second component of Bernoulli's equation at the expense of the third (kinetic) one. To do so they have to have a so-called "high solidity" where turbine blades cover most of the inside flow passage resisting fluid flow and building up the water head. In this case the fluid velocity V falls and the component $V^2/2g$ becomes negligibly small compared to the p/ρ component. That is the reason why the higher water head corresponds to the higher efficiency of the hydraulic turbines, reaching magnitudes close to 90% in some cases.

However, the situation is completely reversed for low, ultra-low or free fluid flows. In these cases the pressure energy component p/ρ almost vanishes and kinetic energy becomes the dominant factor. How would conventional turbines perform in these conditions? They still demonstrate relatively good efficiency because of advanced hydropower technology. But good turbine efficiency using conventional turbines in low-head applications is achieved at the expense of the cost of power as one can see from Exhibit 1.1.

In 1931 Darrieus patented his new reaction turbine which, in contrast to the commonly used wheel-type turbines, has a barreled shape with a number of straight or curved-in-plane airfoil blades and a shaft that is perpendicular to the fluid flow. The Darrieus turbine was greeted enthusiastically by engineers and scientists in both wind and hydro power industries because of its simplicity and because the turbine allowed high speed to develop in slow fluids, maintaining a large passage area without substantially increasing its diameter. However, in spite of numerous intensive attempts over decades to utilize the Darrieus rotor, it has not had wide practical application mostly due to the pulsating nature of its rotation and its relatively low efficiency. Fatigue failure of blades is common in this turbine because of inherent vibration. It also has a problem of self-starting at low rotational speed due to its straight blades which change angles of attack traveling along a circular path.

HELICAL TURBINE

The new helical turbine (see Exhibit 1.2) which was developed in 1994-95, has all the advantages of the Darrieus turbine without its disadvantages, that is, allowing a large mass of slow water to flow through, capturing its $V^2/2g$ kinetic energy and utilizing a very simple rotor as a major factor of the turbine's low cost. The helical arrangement of the rotor blades dramatically changes the performance of the Darrieus-type turbine resulting in the following characteristics:

- high speed uniform spinning in relatively slow fluid flow (low pressure fluids),
- unidirectional rotation in reversible fluid currents,
- high efficiency,
- no fluctuation in torque,
- no visible signs of cavitation in water for high rotating speed,
- self-starting in slow waters or winds.

More than 180 measurements made during 1994-95 in tests of 20 small 3.5 inch (9 cm) diameter helical turbine models demonstrated up to 95% greater power and about 50% higher speed than the comparable Darrieus rotor. Based on these experiments we could expect that a scaled-up helical turbine would have good efficiency without oscillation in both straight and reverse low-head water flows. A scaled-up demonstration project testing a turbine, this time with a triple-helix turbine (see Exhibit 1.3), in fact validated these expectations.

The turbine was thoroughly tested from June-August 1996 in the Cape Cod Canal in Massachusetts. The canal itself presents a unique set of parameters attractive for the installation of a tidal energy demonstration project. The tidal current there reverses four times a day and is very turbulent, with treacherous eddies, vortices, and floating seaweed that somewhat complicates testing of the turbines. The maximum water speed measured at the site was about 5.5 f/s. Turbines developed a firm unidirectional rotation when water velocity was about 1.6-1.8 f/s (about one knot). The overall view of the test site is shown in Exhibit 1.4.

The three-blade turbine was mounted underneath a small raft (10x8x2 feet) and reinforced by steel braces. The 1.25-inch diameter shaft of the turbine was extended upward through the raft providing data on the turbine's torque and speed. Turbine dimensions are:

diameter = 24 inches;
height (length) = 34 inches;
blade profile = NACA-0020 with a 7 inch chord.

In spite of the natural difficulties encountered, the turbine demonstrated quite good performance with a power coefficient of about 35% for maximum loading and water velocities of about 5 f/s. Turbine rotational speed under load was about 100 RPM. Velocities higher than 5 f/s were rarely observed at the site. Concerning the seaweed factor, we note that although the high speed of turbine rotation protected it from an accumulation of seaweed, a substantial amount of the grass did build up on elements of the supporting frame. We also observed substantial corrosion of aluminum parts of the frame at their contact points with steel parts due to electrolysis in the salt water. All of these factors should be taken into account for future designs.

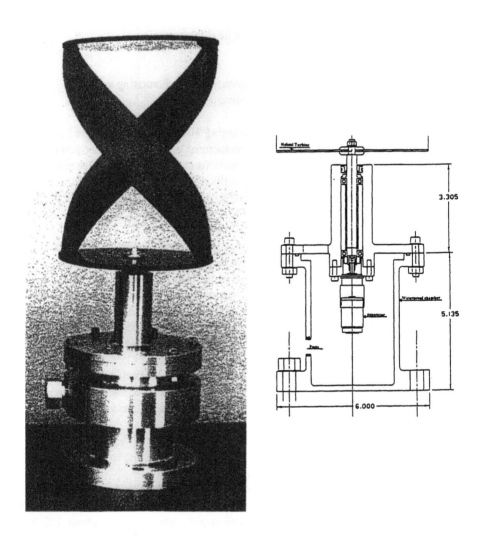

Ex. 1.2 Helical turbine with electric generator in water-sealed chamber

Three blade helical turbine features a 20-inch diameter. The fiberglass turbine with steel shaft weighs 50 pounds.

Exhibit 1.3 Turbine tested in the Cape Cod Canal

The following specific characteristics of the turbine make it different from other hydraulic machines. The turbine consists of one or more long helical blades that run along an imaginary cylindrical surface of rotation like a screw thread (Exhibit 1.5). The helical airfoil blades provide a reaction thrust perpendicular to the leading edges of the blades that can pull them faster than the fluid flow itself. The high speed with no vibration of the helical turbine in a relatively slow fluid, along with structural simplicity, is the key to its efficiency. (A more detailed technical description of the turbine is provided in the literature sources suggested in Notes 2, 3, and 4 at the end of this chapter.)

The helical turbine allows reduction of its diameter while simultaneously increasing its length with no power loss. This is an interesting and advantageous feature of the turbine which can affect the traditional approach to design of a power house, as is shown in the following.

Any high-speed hydraulic or gas turbine has a strength limit that corresponds to its maximum power output. Since the linear velocity reaches its maximum on the periphery of a rotating wheel, it is clear that the major portion of the torque is developed by the parts of the turbine farthest from its center of rotation. This is one of the reasons why engineers try to design turbines of maximum diameter with numerous short blades positioned along the outside boundary of the wheel. The bigger the turbine diameter, the greater its power output for the same angular speed ω and the same shape and sizes of the blades.

However, there are limits to how much the diameter of the turbine can be increased due to the possibility of structural failure caused by centrifugal forces and other dynamical effects.

From this point of view, the helical turbine has a unique advantage since its length L is not limited by centrifugal forces and can be as long as desired. The product DL is approximately equal to the cross-sectional area of the fluid flowing through the helical turbine.

It is apparent from this discussion that the helical turbine allows a new approach to the design of hydraulic or gas power systems using prefabricated turbine modules. Indeed, if a helical turbine module is designed for optimal airfoil and for optimal ω, D and L, the entire power system can be assembled from such modules in either way, shown in Exhibit 1.6. It will be simpler to construct and exploit a power station using multiple helical module turbines because a common shaft can be used for a number of turbines and a single electric generator. It should be noted that the shaft will not be needed in most applications where turbines can be bolted to each other. In this case the torque from each turbine will be transmitted directly to the adjacent turbine by connectors on side discs. The modular design of the turbine runner will simplify the maintenance of the station and reduce cost of its construction.

In November 1997 the scaled-up triple-helix turbine shown in Exhibit 1.7 was thoroughly tested at the University of Michigan Hydrodynamics Laboratory by the Allied Signal Aerospace Company, which also supports the research project. The turbine was mounted vertically underneath a rolling bridge and then pulled with different speeds through the fresh water of the 360-foot long canal. The range of velocities during the test was from 1-10 ft/sec. No shrouding or ducting was used to improve inflow or outflow of the water through the turbine. The objective of the test was to observe performance of the free helical rotor in the natural water streams.

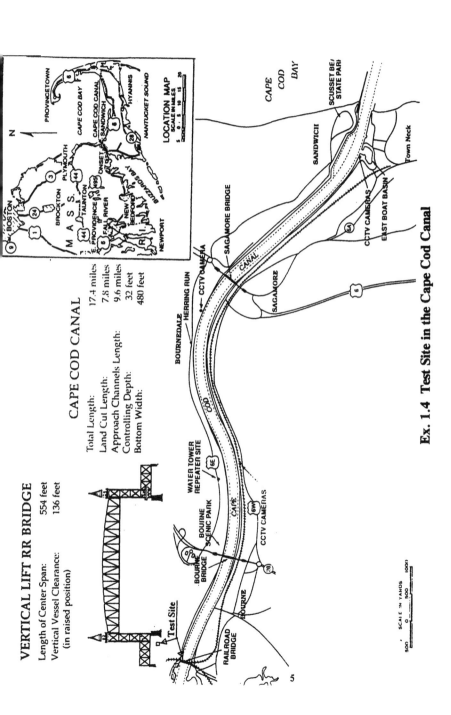

VERTICAL LIFT RR BRIDGE

Length of Center Span: 554 feet
Vertical Vessel Clearance: 136 feet
(in raised position)

CAPE COD CANAL

Total Length:	17.4 miles
Land Cut Length:	7.8 miles
Approach Channels Length:	9.6 miles
Controlling Depth:	32 feet
Bottom Width:	480 feet

Ex. 1.4 Test Site in the Cape Cod Canal

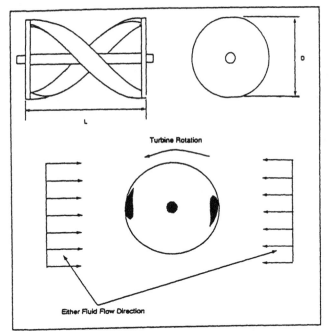

The helical turbine consists of long blades running along a cylindrical surface like a screw thread. The blades can provide a reaction thrust from flows in either direction without significant vibration. The design of the turbine allows the engineer to reduce the relative diameter (D) of the machine while simultaneously increasing its length (L) with no power losses, providing important benefits in hydro project design.

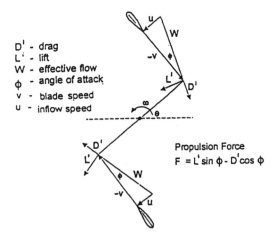

D' - drag
L' - lift
W - effective flow
φ - angle of attack
v - blade speed
u - inflow speed

Propulsion Force

$$F = L' \sin \phi - D' \cos \phi$$

Ex. 1.5 Double-helix rotor

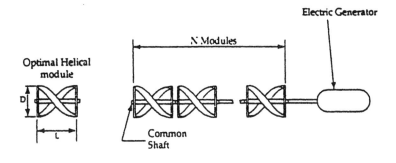

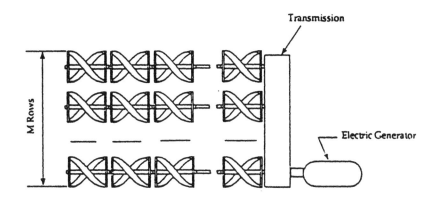

Ex. 1.6 Various turbine assemblies

The flexibility and adaptability of the helical turbine could allow multiple units to be configured in sequence at a small hydropower site. As shown in this exhibit, the turbines could use common shafts and a single electric generator. In ultra low-head applications, the shaft-connected rotors could operate free in the water. Where head or other flow conditions required, simple casings could be designed. The modular design of the turbine runner would simplify the maintenance.

**Ex. 1.7 Triple-helix turbine 40 dia. x 33" tested at
University of Michigan**

Three basic characteristics were measured and documented during the test: relative velocity V of the current in feet per second (i.e., the speed of the rolling bridge); torque T developed by the turbine shaft in pounds per inch; and angular velocity ω of the turbine in rpm's. Then turbine power P_t is calculated as $P_t = T\omega$, and turbine efficiency (power coefficient) is calculated as $\eta = P_t/P_W$, where $_W = 0.5 \rho A V^3$ is the power of the water flow that corresponds to the cross-sectional area A of the turbine. In this case A = 40 x 33 in^2. Torques T were registered by the torquemeter attached to the turbine shaft. The torquemeter was equipped with a hydraulic brake device which allowed for changing the loading on the turbine. During each run with fixed water velocity V, the torque T gradually increased until reaching the maximum magnitude T_{max} which the turbine can carry without stopping. This instant corresponds also to the point of maximum turbine power, P_{tmax}, at minimum angular velocity ρ.

Exhibits 1.8 and 1.9 demonstrate major results of the test. The data shown are calculated for maximum turbine power at each magnitude of the water velocity V. As one can see from Exhibit <5>, the turbine develops a stable efficiency, around 35%, at all water velocities. Starting with a firm rotation at water flow V of about one knot, the turbine increases its power in proportion to the water velocity cubed. No oscillation or vibration of the turbine was observed during the test.

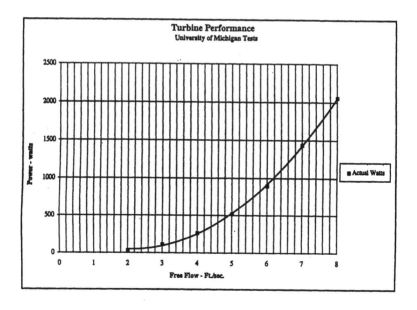

Ex. 1.8 Output power vs. water flow

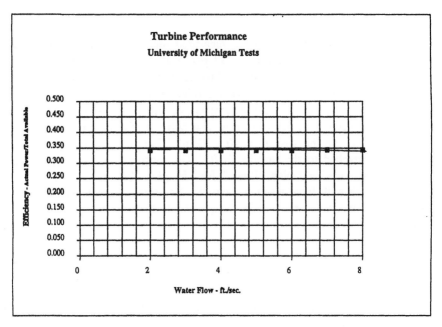

Ex. 1.9 Turbine efficiency

Tip ratio V_t/V, where V_t is the linear velocity of the turbine blade, is dependent on the magnitude of the loading torque. For the maximum torque applied to the shaft, the tip ratio was quite stable, in the range of 2.0-2.2.

OCEAN POWER FARMS

Helical turbines can be used as the key power modules in the design of ocean power farms for harvesting the energy of ocean streams. Such farms, if built in major ocean streams, such as the Gulf Stream near the North American continent or the Kuroshiwo Current near Japan, can produce hundreds or even thousands of megawatts of electric power. Moreover, once installed in the ocean stream the power farm can be expanded in the future as much as desired since the energy potential of the ocean streams is greater than any imaginable requirements of mankind. For example, the mass of water carried by the Gulf Stream in the Atlantic Ocean at 38° north latitude is 82 million m^3/s, which is many times greater than the water flow of all the earth's rivers together.

The following is a conceptual approach to the design of a straight or reversible (tidal) ocean stream power farm using helical turbines to extract energy from the water current. Turbines will be positioned vertically. This makes power production of the entire farm independent of the direction of the water stream since the helical turbines are unidirectional rotation machines. Such a design is especially advantageous in reversible tidal streams or streams that change direction depending on which way the winds blow, for example, during monsoons.

We consider the three-blade helical turbine shown in Exhibit 1.7 as the optimal power module for this project. As mentioned, this turbine has demonstrated 35% efficiency in free water flow.

Since the overall dimensions of a stream farm -- namely its length, width, and depth -- depend on the designed power capacity, it is convenient to choose a reasonably small modular farm and then to increase the project capacity if necessary by adding more modules. Such a floating modular farm schematic, shown in Exhibits 1.10 and 1.11, consists of the following principal mechanical and structural components:

- helical turbines;

- electric generators, each of them designed to pick up power from a vertical assembly of 16 turbines mounted one upon another;

- a floating frame constructed from prefabricated longitudinal, lateral and vertical elements, which could be built from metallic or plastic tubings. The frame performs two functions:
  1. As a structural system, which provides the integrity and strength of the farm, and
  2. As a pontoon, which maintains flotation of the farm at the designated depth level;

- Anchors that secure the position of the farm against the ocean current pressure and the hydrostatic lifting force.

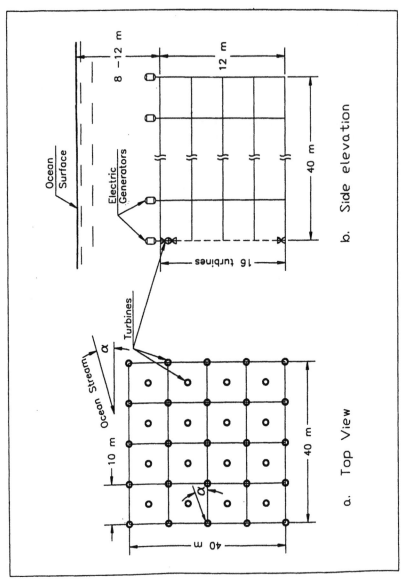

Ex. 1.10 Stream farm schematic

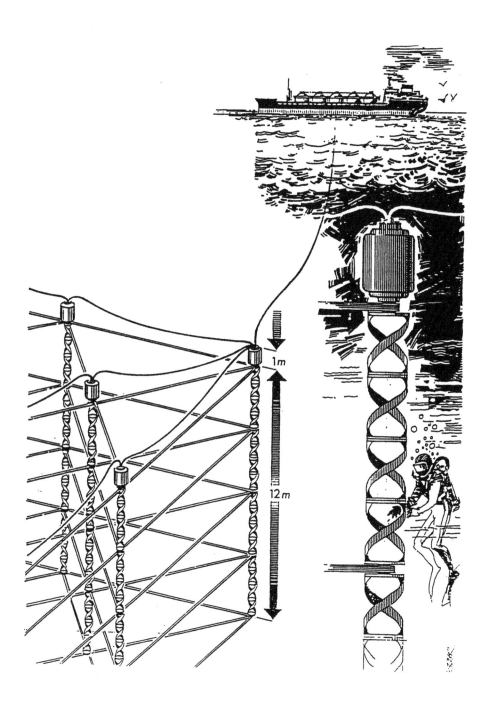

Ex. 1.11 Detail of the farm

Exhibit 1.10a shows the top view of the farm with dimensions of 40 x 40 meters between axes of the extreme turbine rows. The side elevation is shown in Exhibit 1.11b. The farm consists of five rows of vertical turbine assemblies, each with 16 modular turbines, mounted on a common shaft with one electric generator on the top. Every other inner row contains four turbine assemblies. The span of 10 meters between adjacent turbines along orthogonal axes is chosen so as to be big enough to reduce interference of the turbines with each other, and to be reasonably small enough to avoid unjustifiable expansion of the structure. To minimize obstruction of the turbines against water flow through the farm, it must be turned horizontally to make a small angle α with respect to the ocean stream (see Exhibits 1.10 and 1.12). In this case all the turbines would fully open to the water pressure from the stream. The smallest offset angle is $\alpha = 6°$ in our case, as in the diagram of Exhibit 1.12. Also, to protect the farm against the damaging effect of storm waves, the entire structure can be positioned 10-15 meters or even deeper below the ocean surface.

To estimate the overall power capacity and cost of the modular stream farm, let us consider the specific conditions of the Gulf Stream where the water velocity is V = 2.5 m/s, that is, a little less than 5 knots. Taking the cross-section frontal area of the turbine A = 0.865m^2, we calculate power of the free water flow through such a section as

$$P_W = 1/2 \, \rho A V^3 = 6.87 \text{ kW}$$

where density ρ of the sea water is about 1015 kg/m^3. The power of one turbine with 35% efficiency is calculated as:

$$P_t = 0.35 P_W = 2.4 \text{ kW}.$$

Because the modular farm contains 16 x 41 = 656 turbines, the total power of the module P_m is:

$$P_m = 1.6 \text{ MW}.$$

Taking into account the combined efficiency of electric generators (including losses in electric circuits) as 85% we can obtain the modular farm power output:

$$P_m = 1.36 \text{ MW} = 1,360 \text{ kW}.$$

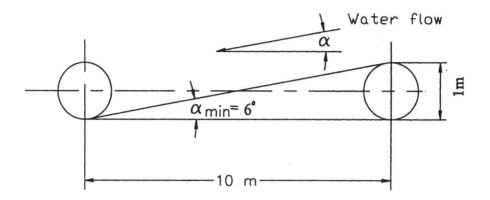

Ex. 1.12 Angular offset of farm position

Approximate cost estimation for elements of the farm includes:

Elements Needed	Cost
1. One turbine as shown in Ex. 1.3	$700
2. 38 kW power waterproofed electric generator for assembly of 16 turbines	$2,000
3. 40x40x12m frame from prefabricated tubings with some steps and small platforms	$1,000,000
4. Anchors	$300,000
5. Electric cables, connectors etc.	$100,000
6. Other mechanical and electrical supplies	$100,000
7. Transportation, installation under water and levelling of the farm	$1,000,000

Thus, the total cost of the modular farm would be:

$$C = \$700 \times 656 + \$2,000 \times 41 + \$1,000,000 + \$300,000 + \$100,000 + 100,000 + \$1,000,000 = \$3,040,000$$

This converts into a unit cost of installed power:

$$C_u = \$3,040,000/1,360 \text{ kW} = \$2,235/\text{kW}.$$

The cost of operating the farm was not included in the above analysis. This cost would depend on the specific requirements for the technological operation of the farm, i.e., where and how the electric power generated would be used, etc.

For comparison, the unit costs for the construction of other well-developed power technologies are summarized in the following table:

Unit cost of installed power in $/kW		
	Technology	**Cost**
1.	Nuclear	$2,500
2.	Coal Fired	$1,500
3.	Oil Steam Plant	$1,300
4.	Conventional Hydro (site dependent)	$1,500-2,500
5.	Solar	$5,000-8,000

Thus our calculation of $2,235/kW for an ocean power farm is compatible with prices of other power technologies.

However, the helical turbines do not need any fuel for their operation and do not pollute the water or air. From this point of view the ocean power farm can be compared with solar power systems, which are substantially more expensive.

We believe that a reasonable field size for a power farm under Gulf Stream conditions should be 400x400m containing 100 modular farm units as discussed. Such a stream farm would continuously generate 136 MW power, and its construction cost would be around C = $300 million. Turbines are now in production at Allied Signal Aerospace Company, and we hope by 2001 to have such a farm in operation off the coast of Florida where the stream velocity is around 8 ft/sec (5 knots) with a top supply of 30 megawatts.

The final issue to be addressed is how and where to use the electric power generated by the ocean stream farms. We envision two obvious options. The first is to transmit electric power from the farm to the mainland for further conventional use. This can be done by means of an electric cable on the ocean floor when such a project becomes technically and economically feasible.

The second option is to utilize the farm power *in situ* for year-round production of hydrogen fuel by electrolysis of the ocean waters. Liquefied or stored by any other method, hydrogen can then be transported everywhere to be used either instead of gasoline in internal combustion engines or in fuel-cell motors. We studied such an option in our earlier projects and found it feasible if cheap electric power is available in large quantities.[5] For *in situ* production of hydrogen fuel, a well-equipped floating electro-chemical factory should be positioned next to the stream farm. This factory would use electric power from the farm to resolve the ocean water into hydrogen, oxygen, and some other chemical byproducts and store them for further transportation. Obsolete tankers or other large naval vessels can be converted into such factories. Exhibit 1.13 illustrates the ocean power project described, including both the stream farm and the floating hydrogen production factory on the open ocean.

As mentioned, the power capacity of the stream farm can easily be increased by adding more helical turbine modules. One can visualize mega-power farms such as those in Exhibit 1.13 in the Gulf Stream or the Kuroshiwo current able in the near future to generate thousands of megawatts of electric power. Such ocean power farms with helical turbines can be of particular value for future floating cities, which have been projected for overpopulated countries such as Japan. Exhibit 1.14 demonstrates the possibility of using the power farm as a permanent power supply source for such a floating city.

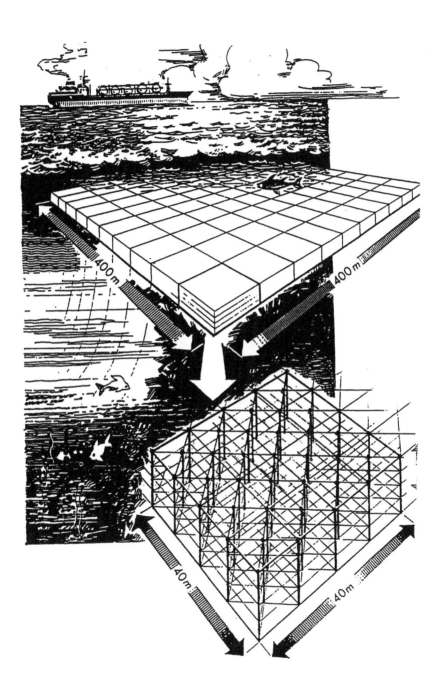

Ex. 1.13 Project ocean power farm

Ex. 1.14 Helical turbines as a power source for marine cities

MINI POWER STATIONS

The new helical turbine technology opens up prospects for designing mega-projects to harness the limitless energy of ocean streams and tides.

However, there are thousands of sea and river sites in the world where small power stations can be constructed to supply electric power to local consumers. Such mini power stations of 2-5 kilowatts can be easily built and utilized by small communities or even individual households located near direct or reversible water streams. Assembly of a few helical turbines, such as shown in Exhibit 1.15, can supply a small consumer with permanent electric power from the renewable hydro energy source without construction of any dam. In this case the global macro-energy objective would be solved by means of numerous mini power installations, similar to those discussed previously.

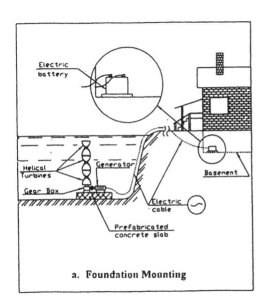

a. Foundation Mounting

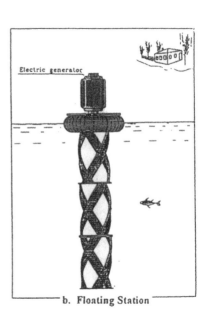

b. Floating Station

Ex. 1.15 Mini power stations

APPLICATIONS FOR OCEAN WAVES

The kinetic energy of ocean waves also is an untapped, renewable energy source. The absence of reliable and low-cost technologies to convert this energy to useable electrical energy is the major barrier to exploitation of this abundant energy source. However, the helical turbine offers a solution to this difficult problem either. Exhibit 1.16 demonstrates one of the possible applications of helical turbines in this case.

Power output from the helical turbine is proportional to the flow rate cubed. While open ocean currents in many locations are below the power threshold of the turbine, wave-generated flows are typically in the range of 2-10ft/sec with direction reversals every 5-12 seconds. Flows of this magnitude are capable of producing significant power with even relatively small turbines.

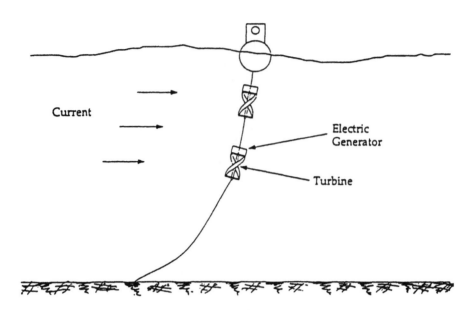

Ex. 1.16 Helical turbine as a continuous energy source in water currents

Deep water surface moorings used for oceanographic research typically use 10-foot diameter discus or hemispherical buoys with 8-20,000 pounds of buoyancy. They are moored with a combination of wire rope and nylon line. The nylon provides compliance for waves, current and winds. As the surface buoy moves with the seas, the motion is converted to primarily vertical oscillations on the mooring line. In the wire section of the mooring, the flow rates produced are in the 2-10 ft/sec range depending on wave height and period. A helical turbine with its rotational axis perpendicular to the mooring line will rectify the oscillatory flow of the line and produce power proportional to the cube of the instantaneous velocity. Multiple turbines can be deployed on a mooring to provide power to various sensors, sound projectors, or other energy intensive instruments.

Surface moorings are not the only means of extracting wave power with the helical turbine. A sub-surface mooring with the buoyancy element positioned near the surface experiences substantial forcing at wave frequencies due to fluctuating pressure and drag at the subsurface buoy. With a long steel or nylon mooring line, these forces cause the mooring to oscillate along the vertical axes at its natural period. The amplitude of the oscillation is a function of the compliance of the mooring material and the frequency of oscillation is a function of the mooring design. A properly tuned mooring can be made to oscillate substantially at typical wave periods and thereby provide significant vertical velocities that can be harnessed by the turbine.

Finally, in shallow water a vertical mounted turbine near the bottom would be exposed to substantial horizontal oscillary flows generated by waves at the surface. Depending on the water depth and the wave period, these horizontal flows may be as large as the vertical flows seen on deep water moorings. Thus, we can identify three potential wave power generating systems: surface moorings, sub-surface moorings, and bottom moorings (shallow water only).

HELICAL TURBINES FOR WATER PUMPING

A substantial portion of the electric power in many developing countries is used to pump water from rivers into irrigation supply systems. This is a common situation in such countries as India, Pakistan, and Egypt where there is intensive use of irrigation canals in agricultural technology. The helical turbine can be helpful in these cases because it can directly convert kinetic energy of river water flow into mechanical energy to operate water pumps.

Exhibit 1.17 demonstrates components of a helical turbine-water pump assembly designed to supply water from rivers directly into an irrigation system. No intermediate generating of electric power is needed in such an installation. One or more turbines can be mounted in the chain depending on turbine capacities, water pumping rate and water velocity in the river. Assembling more turbines on the same shaft across the water flow would increase overall power of the pump station accordingly to the water demands.

The helical turbine-pump system, once installed in the river, becomes, in certain respects, a "perpetual motion" (excluding maintenance) device supplying water as long as the river keeps flowing. This system should find wide application in developing countries due to its obvious simplicity and low cost.

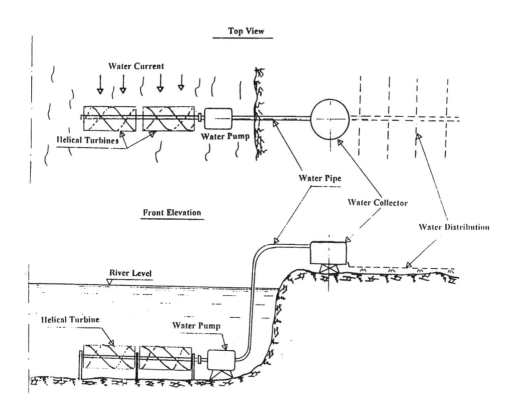

Ex. 1.17 Helical Turbine Water Pump system for irrigation

WIND FARMS WITH HELICAL TURBINES

The hydraulic helical turbine is recognized by hydro power corporations and individuals around the world as an efficient new apparatus to harness hydro energy from ocean streams, tidal estuaries, low-head rivers and canals. The turbine performance has been proven in both laboratory and field testing. The performance characteristics of the turbine, which include independence of the direction of the fluid flow, pulsation-free rotation over the entire cycle, and high efficiency, make the helical turbine an excellent candidate for application in wind power systems as well. This turbine should demonstrate about the same characteristics in wind-harnessing installations as it does in water. However, its design for wind requires a different approach for optimization due to different air density, viscosity, exploitation conditions, and velocity of rotation.

We suggest a wind farms design using multiple standard-sized helical turbines in two different options as follows:

- a. Horizontal assembly of the turbines (see Exhibit 1.18)
- b. Vertical assembly of the turbines (see Exhibit 1.19)

The design in Exhibit 1.18 is more power-efficient than the one in Exhibit 1.19 because all the turbines are elevated high above the ground where winds are usually much stronger than they are near the land surface. The disadvantage of such a system is its dependence on the direction of the wind, in other words, from the angle between the direction of wind and the axis of turbine rotation. For Exhibit 1.18, the maximum power would be generated if wind flow is perpendicular to the turbine shafts. If the wind can change its direction with respect to the turbine axis, the design in Exhibit 1.19 should be used.

Compared with a conventional wind power system using a high tower with a single large-diameter propeller, the helical turbine has the following advantages for wind applications:

- It rotates in the same direction even if the wind changes its directions,
- It does not pulsate in constant wind velocity in contrast to the Darrieus wind rotor,
- Maintenance of the wind farm is simple because any turbine in the system can be easily removed and replaced,
- The same size relatively small standard helical turbines can be used for wind farms of different power capacities.

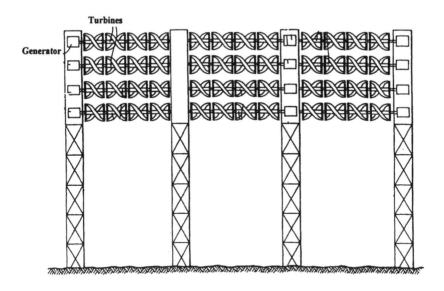

Ex. 1.18 Wind farm with horizontal helical turbines

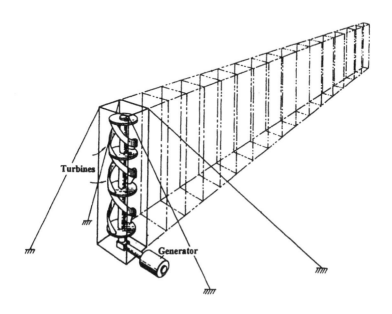

Ex. 1.19 Wind farm with vertical helical turbines

MATHEMATICAL MODEL FOR DESIGN AND OPTIMIZATION OF THE HELICAL HYDRAULIC TURBINE

Let us consider a stationary turbine with a helical blade that runs along the thread line AME on the surface of a cylinder of height **L** and radius **R** (see Exhibit 1-20). The equation of the helix is defined by

$$x = R \cos \varphi, \qquad y = R \sin \varphi, \qquad z = R\varphi \tan \delta \qquad (1)$$

where x, y and z are coordinates of the point **M** of the helix, and δ is the angle of inclination of the blade to the **XOY** plane.

We assume, as an approximation, that the cross-section of the blade has the shape of an infinitely thin rectangle with its length equal to the chord **b** of the blade's airfoil. This does not change the proportion between lifts and drags after resolution of the reaction force **F** which can be calculated as

$$F = k_o A V_w^2 \qquad (2)$$

where:

k_0 = constant. In this case k_0 is set to about 1.2 ρ (ρ - water density),

A = projection of the frontal area of the blade on the plane perpendicular to the water flow,

V_w = water velocity.

Designating a segment of the curve AME by ξ we obtain

$$d\xi = R(1 + q^2)^{1/2}d\varphi = \frac{R}{\cos \delta}d\varphi, \quad \text{where } q = \tan\delta \qquad (3)$$

From (3):

$$\xi = R(1 + q^2)^{1/2}\varphi = \frac{R}{\cos\delta}\varphi, \qquad (4)$$

$$\varphi = \frac{\cos \delta}{R}\xi = \frac{\xi}{R\sqrt{1+q^2}}.$$

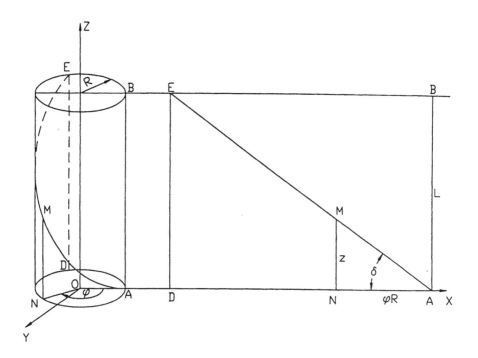

Ex. 1.20 Development of the blade line on vertical plane

On the other hand

$$z^2 + (R\varphi)^2 = \xi^2$$

$$L^2 + (R\varphi_0) = 1^2$$

where:

1 = the length of the blade,

φ_0 = φ_{max} is the angle of twist of the entire blade.

Let's designate the angle of attack by α. Then, at any point M

$$\sin \alpha = \cos \varphi \quad \text{and} \quad \cos \alpha = \sin \varphi. \tag{5}$$

The torque T is obtained from the equation

$$T = FR \cos \alpha. \tag{6}$$

Substituting (2) in (6) we obtain torque ΔT for a small area of the blade $\Delta A = b\Delta\xi\sin \alpha$:

$$\Delta T = k_0 b \Delta \xi V_w^2 R \sin \alpha \cos \alpha \tag{7}$$

where b is the airfoil cord, and $\Delta\xi$ is a small segment of the blade along the AME line.

When $\Delta\xi\_0$

$$T = k_1 \int_0^1 \sin\alpha\cos\alpha \, d\xi \qquad (8)$$

where

$$k_1 = k_0 \, b \, V_w^2 \, R \qquad (9)$$

From (4), (5) and (8)

$$T = k_1 R\sqrt{1+q^2} \int_0^{\varphi_0} \cos\varphi \, \sin\varphi \, d\varphi = k_1 R\frac{\sqrt{1+q^2}}{2} \sin^2\varphi_0 \qquad (10)$$

Since $q = \tan\delta = \dfrac{L}{\varphi_0 R}$ the total starting torque developed by the blade in the water flow V_w can be expressed as

$$T = k_2\sqrt{1+q^2} \sin^2\left(\frac{L}{Rq}\right) \qquad (11)$$

where

$$k_2 = \frac{1}{2} k_1 R \qquad (12)$$

Or, in dimensionless representation:

$$T_1 = \frac{T(q)}{k_2} = \sqrt{1+q^2} \sin^2\left(\frac{L}{Rq}\right) \qquad (13)$$

Exhibit 1.21 represents the torque T_1 as a function of angle of blade inclination δ and the ratio L/R of the turbine height to its radius. It is remarkable that the torques reach their maximums for different δ with changing L/R. The angle δ increases with increasing ratio L/R. This means that for the constant R the higher turbine has to have its helical blades closer to the vertical line in order to obtain the maximum torque.

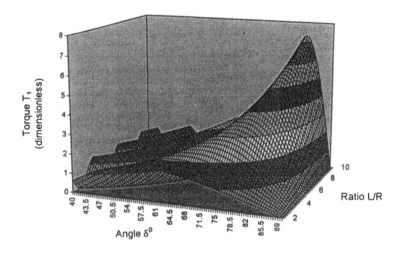

3D Diagram $T_1 = f(L/R, \delta^0)$

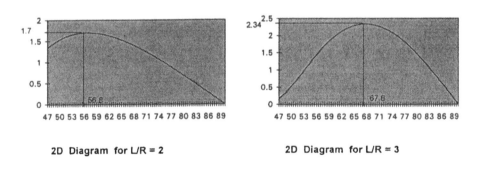

2D Diagram for L/R = 2   2D Diagram for L/R = 3

Ex. 1.21 Torque as a function of angle δ and L/R ratios

COMPARATIVE PERFORMANCE OF HELICAL VERSUS DARRIEUS TURBINES

Since the helical turbine is a modification of the well-known Darrieus rotor, it was possible to compare them in almost identical laboratory tests.

The following same size turbines were thoroughly tested and compared:
- Helical Turbine with 3 blades twisted on 60° angle,
- Darrieus Turbine with 3 straight blades.

Heights of both turbines were 9 inches, the diameter was 8.5 inches. The airfoil cross-sections are NACA-0020. Plastic blades were built at the SLA-190 rapid prototyping machine. Two plastic disks were used on both sides of each turbine for mounting blades and transmitting torques. The objective of the experiment was to compare two turbines but not to optimize them to obtain maximum efficiency.

Each turbine was tested in the water flow with small heads ranging from 1 to 8.5 inches and water velocities from 0.9 to 2.4 ft/s.

The following tables summarize the collected data for the turbines.

A. Helical Turbine

Water Head [in]	Water Power [watts]	Peak Turbine Power [watts]	Peak Efficiency [%]
1.0	3.1	0.5	16.2
2.25	10.5	1.95	18.6
3.0	16.6	3.21	19.3
4.75	30.0	5.85	19.5
6.5	54.8	11.9	21.7

B. Darrieus Turbine

Water Head [in]	Water Power [watts]	Peak Turbine Power [watts]	Peak Efficiency [%]
2.0	7.9	0.8	10.5
3.25	16.1	2.0	12.4
5.0	29.2	4.2	14.3
7.75	57.2	9.7	16.9
8.5	65.1	12.0	18.4

Tables A and B and Exhibits 1.22 and 1.23 reflect comparative characteristics of both turbines including the turbines' power and their efficiency (power coefficients) depending on water heads and water power. As can be seen from these charts the helical turbine demonstrates substantially better performance than the Darrieus rotor in all major characteristics including turbine power, efficiency, and speed of rotation. In many comparisons the helical turbine is in excess of 50% and higher.

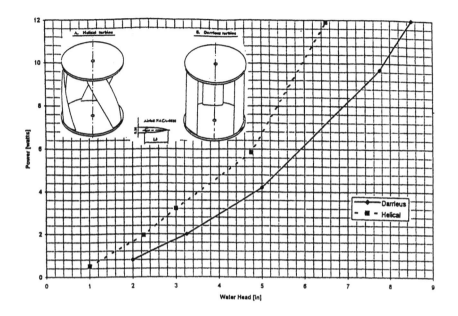

Ex. 1.22 Peak turbine power versus water head

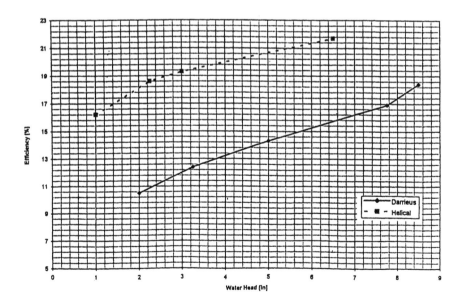

Ex. 1.23 Peak turbine efficiency versus water head

Experiments revealed the remarkable advantage of the helical turbine over the Darrieus rotor, namely, its smaller resistance to the water flow. This resistance depends on the turbine's so called solidity defined as

$$\sigma = \frac{bi}{D}$$

where **b** is chord of the blade section, **i** is number of blades and **D** is turbine diameter. The larger the product **bi**, the greater is the resistance of the turbine to the water flow for the same **D**.

However, this resistance depends also on the turbine's speed of rotation. The higher the turbine speed, the more it obstructs the stream. A very fast rotating turbine in free water with no ducting would practically stop the water flow through it, reducing turbine efficiency to zero.

Thus, turbine resistance is a very important characteristic that can substantially reduce efficiency of turbines in free currents. Most of the water simply avoids the turbine without producing any useful work, when the turbine develops high resistance to the water flow. In our experiments the turbine resistance was evaluated by measuring elevation of the water level (water heads) in front of the rotating turbine. Those observation are reflected in charts of Exhibit 1.24 showing both helical and Darrieus turbines.

As one can see, the Darrieus turbine develops water heads from 30% to 50% higher than the helical turbine in all ranges of their rotating velocities (rpm). For example, the ratio of water heads of Darrieus and Helical turbines for 150 rpm turbine speed without load is about 1.5 (top chart of Exhibit 1.24). This ratio remains the same (1.5) for turbines under maximum load and the same velocity 150 rpm (bottom chart of Exhibit 1.24). So, in both cases the Darrieus turbine exhibited 50% stronger resistance to the water flow than the same size helical turbine.

Another important characteristic of hydraulic turbines which actually triggered the present research is their oscillation and vibration under the load. Oscillation of the turbine causes not only fluctuation of the electric power, but it also leads to the fast failure of mechanical parts and joints in the turbine-generator-transmission chain. From this point of view, the helical turbine demonstrates obvious superiority to the Darrieus-type rotor. While the helical turbine did not show any sign of vibration during testing, the Darrieus turbine oscillated substantially in all the experiments. Pulses from the Darrieus turbine were especially strong when its blades passed the walls of the water channel.

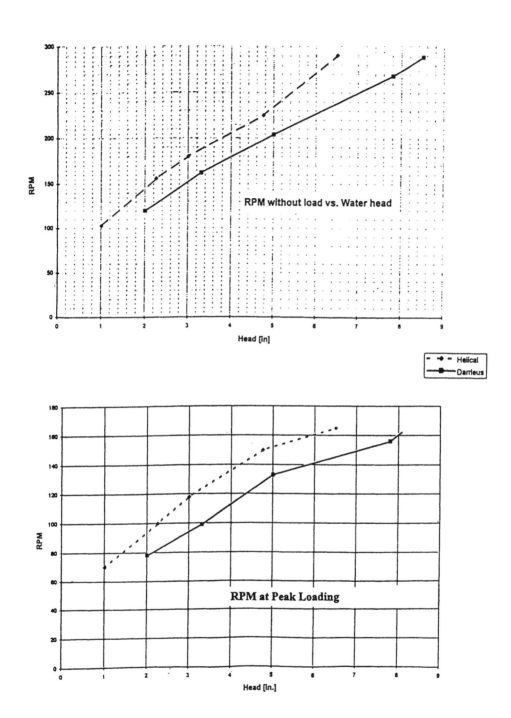

Ex. 1.24 Turbine rotation with and without loading

NOTES

1. Faure, T.D., Pratte, B.D., & Swan, D. "The Darrieus Hydraulic Turbine Model and Field Experiments." Proceedings of the 4th International Symposium on Hydropower Fluid Machinery, ASME, New York, 1986. See also, Takamatsu, Y. et al., "Experimental Studies on a Preferable Blade Profile of Darrieus-type Cross-Flow Water Turbine", *JSME International Journal*, Vol. 34, No. 2, 1991.

2. Gorlov, A.M., "The Helical Turbine: A New Idea for Low-Head Hydro," *Hydro Review*, No. 5, 1995.

3. Gorlov, A.M., Unidirectional Helical Reaction Turbine, U.S. Patent No. 5,451,137, Sept. 19, 1995.

4. Gorlov, A.M. & Rogers, K., "Helical Turbine as Undersea Power Source," *Sea Technology*, December, 1997.

5. Gorlov, A.M., "Hydrogen as an Activating Fuel for Tidal Power Plant," *International Journal of Hydrogen Energy*, Vol. 6, No. 3, 1981.

2

Power from Space

Peter E. Glaser
President, Power From Space Consulting Co., Inc.

INTRODUCTION

For the past three decades one of the international strategic objectives for space missions has been the assessment, development and demonstration of the Solar Power Satellite (SPS)[1] as a non-depletable and ecologically compatible source of energy. The objective of SPS is to meet foreseeable energy demands consistent with sustainable global development and with minimum deleterious effects on the earth's ecology compared to currently used energy sources.[2]

The SPS options, designed to convert solar energy in space and beam the power to receivers on earth, are increasingly recognized internationally as key to raising living standards worldwide. SPS utilizes wireless power transmission (WPT) which heightens its interest because of the growing acceptance of its technical feasibility, economic viability, and ability to meet projected global energy demands in the future.[3]

Currently, the impulsive actions of our species appear to violate life's evolutionary path on this small planet. Now is the time to organize a global consensus on how best to protect the earth's ecology and to apply with prudence the energy emanating from the sun that sustains all life.

RATIONALE FOR POWER FROM SPACE

One-half of the current population will live in cities by 2000. The present migration into cities of 150,000 people per day may increase to 250,000 per day in 25 years. To achieve the goal of higher living standards, increased energy supplies that are compatible with the ecology will be required on a global scale.

Power from space should be an integral part of global industrialization goals to meet these insistent demands for higher living standards of a population, currently estimated by demographers of the UN Population Division to reach nine billion by the mid-21st century.

There are vast fossil fuel energy resources on earth but some are deep in the ocean or so far underground that they are not yet -- and may never be -- economical to recover. Others, like shale oil, are environmentally unacceptable because they require strip mining and huge amounts of water. Development of oil and gas reserves will attempt to meet growing consumption at a time when more people and countries compete for limited and declining resources. The deleterious consequences on the ecology of continued increases in the use of fossil fuels are as yet not fully determined, but indications are that placing ever- increasing reliance on these fuels beyond several decades is not advisable.

There may also be an unprecedented shift in economic strengths during this time period, with several emerging economies growing faster than those of nations with mature economies. The impact of these shifts will result in greatly increased consumption of available fuels. This is projected to escalate the degradation of the Earth's ecology in the next century. In addition, the finite terrestrial energy resources have already, or will in the foreseeable future, pass their peak availability.[4]

Alternative energy development paths will become dominant in the post-2020 period because of the extended project lead times of power plants, refineries, and planned associated energy resource investments. It is estimated that over the next three decades capital requirements in the energy sector will range between $13 to $20 trillion to keep pace with the growing energy use in developing countries. The latter amount approaches the total global economic output.

Currently operating energy conversion systems may not meet future energy demands in view of their technical, economic and societal challenges which include the following:

- resource limitations of fossil fuels
- attempts to stabilize greenhouse gas concentrations in the atmosphere to prevent interference with the global climate system
- efforts to ensure that all inputs and outputs of the energy conversion system will be usable for peaceful purposes only
- the possibility of catastrophic or long-term effects of the energy supply system on public health
- limits of terrestrial renewable energy systems to supply baseload power demands

THE SPS OPTIONS

SPS would orbit the earth to tap the inexhaustible energy of the sun; a lunar SPS could use commodity materials known to exist on the moon for construction. Both of these options would ensure the availability of unfailing, globally distributed energy for future generations, compatible with the ecology of the earth.

The technologies for gaining access to solar energy in space are photovoltaic conversion and WPT to receiving antennas on earth. WPT is based on the discovery of Hertzian waves, on efforts by Nikola Tesla to demonstrate this technology nearly a 100 years ago, and on the work of William C. Brown, retired from the Raytheon Company, who is credited with major contributions to the development of WPT technology. The current widespread uses of WPT in the microwave spectrum for industrial and domestic applications have already demonstrated the advantages of the applicable WPT technologies which meet international exposure standards established to ensure the health and safety of the public.

Key demonstrations of WPT have been performed during the past twenty years by NASA, the Canadian Department of Communications, the Institute of Space and Astronautical Science, universities and industry in Japan, as well as several universities and research institutes in China, Europe, India, Russia, and Ukraine.

The first international meeting on SPS, organized by the International Microwave Power Institute, was held in 1970 in The Netherlands. Frequent meetings were held during the NASA and Department of Energy Satellite Power System Concept Development and Evaluation Program, at International Astronautical Congresses, and in conjunction with national and international professional society meetings.

Recent conferences devoted to space power systems were held at WPT '91 in Paris, WPT '95 in Kobe, Japan, and WPT '97 in Montreal, Canada. As a result of international contributions to this field, an extensive literature on SPS and space power systems is available.

ASSESSMENT OF TECHNICAL, ECONOMIC, AND SOCIETAL ISSUES

The results of assessments by NASA and the Department of Energy performed from 1975 to 1980 concluded: "No single constraint was identified which would preclude the development of the SPS option for either technical, economic, environmental, or societal reasons."

Although SPS development and construction represent a major challenge to industry, there is growing confidence that given the resources to be expended for conventional power plant construction projects globally (up to $20 trillion by 2025), a global system could begin operation by that time and be competitive when several, widely recognized environmental costs are accounted for.

An approach to developing SPS can be devised that identifies the required generic technologies and their application to specific projects. The "terracing" of SPS projects would reduce the challenges typically associated with macro-engineering projects, including control of the project, effects of technical uncertainties on feasibility, maintenance of investor confidence in the project, effective reduction of potentially negative environmental impacts at all stages of the project, and the early identification of any obstacles that would affect schedule and cost.

The increasing capabilities needed to plan, control, and manage large and complex space projects, such as the international space station, high-performance and lower-cost space transportation systems, and global communication satellite networks, will be applicable to SPS development, construction, and operation.

A variety of development projects completed at successive "terrace levels" of increasing scope and complexity may include point-to-point WPT over limited distances on earth, high-altitude long-endurance aircraft and dirigibles,[5] and power relay satellites.[6] The lessons learned will apply to SPS in earth orbits and on the lunar surface.[7] These projects will be essential to achieving reliable and effective operations, selecting the most appropriate technologies, complying with applicable standards, regulations and laws protecting the public and the environment, and obtaining the required investment funds from public and private sources. The "terracing" approach would increase the confidence of the public and of potential investors that these systems will meet agreed-upon performance requirements.

SPS will have to be designed and operated in compliance with the existing international legal and regulatory framework.[8] They will have to comply with microwave frequency assignments, orbit position assignments, controlled debris generation in orbit, and peaceful operations -- and be in compliance with the regulations of the International Telecommunications Union, a specialized agency of the United Nations established by treaty between 163 member nations.

Furthermore, the United Nations space law has become "the supreme law of the land" through treaties ratified by member states. A major provision requires that space activities be carried out for the benefit of all states. The implementation of SPS projects will be eased as applicable technologies continue to be evaluated, developed, and demonstrated by the growing number of groups at universities, government agencies, and industry in many countries.[9] Although the organization of an international entity to supply power globally is still in the future, consideration is being given to such a possibility because of the global scale encompassed by SPS.

Electric utilities can assess, as part of their planning efforts, how and when the availability of power from space would influence future expansion plans. These efforts are vital in view of the challenges posed by the inevitable increase in demands to supply electric power for sustainable development in an urbanizing world.

Societal acceptance of SPS will be of crucial importance. More information on SPS is becoming available internationally and is accessible to the public. This will enable judgments to be made regarding the significance of this option which could meet future global power requirements without incurring the ecological risks associated with other currently known power generation options.

Providing the required globally distributed power, SPS may also enable the creation of the "hydrogen economy" based on the electrolysis of water. At some future time the power generated by SPS may also permit powerful lasers to change the trajectory of comets and asteroids to avoid their potential impacts with earth.

CONCLUSION

The goal to meet near-term pressing economic development requirements should be pursued in parallel with planning and implementing a sustainable future for humanity. Meeting this goal will require effective management of both terrestrial and extraterrestrial energy sources on an unprecedented scale to meet projected energy demands.

Challenges to the development of extraterrestrial energy and materials resources are formidable. International efforts are already underway and will be required over a period of decades to make the transition from current to 21st century energy production methods.

The interdependence of the earth's ecology is being threatened by the impulsive actions of humans, if past actions are a prologue to the expanded uses of present energy sources. While parts of the solution to meet global energy needs may already exist, all current energy sources have specific drawbacks that limit their expansion on a scale that is now seen as essential. Gaining access to the inexhaustible energy of the sun will ensure that all foreseeable global demands can be met.

Significant changes in the global economy favor future SPS macro- engineering projects:

- Commercial markets for the space industry have come of age. This year, commercial revenues exceeded government revenues in the U.S. for the first time. Billions of dollars are being raised for new space ventures.

- Private sector capitalization for SPS projects is now more viable, especially given the widely recognized need for renewable energy sources and continuing technological advances applicable to SPS.

- Initial investments would approach those of telecommunication satellites. Once the SPS technologies have been demonstrated in pilot projects, SPS could start generating power after five years.

- This investment and development stage (10-15 years) would require the coordinated efforts of government space agencies, aerospace companies, global installations by A&E firms, power utilities, and cooperative international efforts.

Now is the time to select the most desirable power generation options including terrestrial renewable energy sources and SPS before commitments are made that would result in unsustainable global development. WPT to access power from terrestrial renewable sources and from solar energy in space will ensure that the benefits of placing increasing reliance on solar energy will sustain life on earth.

"It may be that in the future, man will be indifferent to stars except as spectacles, but if (and this seems more probable) energy is still needed, the stars cannot be allowed to continue in their old way, but will be turned into efficient heat engines.[10]

NOTES

1. Glaser, P.E., "Power from the Sun: Its Future," *Science*, Vol. 22 November 1968, pp. 857-861.

2. Subcommittee on Global Change Research, National Science and Technology Council, "Our Changing Planet - The FY 1998 U.S. Global Change Research Program". Office of Science and Technology Policy, Executive Office of the President, Washington, DC.

3. Mankins, J.C., "Space Solar Power: A Fresh Look at the Feasibility of Generating Solar Power in Space for Use on Earth," NASA, April 4, 1997, Report No.SAIC-97/1005.

4. Duncan, R.C., "The World Petroleum Life-Cycle: Encircling The Production Peak," *Space Manufacturing 11, The Challenge of Space: Past and Future.* Space Studies Institute, Princeton, NJ, 1997, p. 275.

5. De Laurier et al, "Research on the Technology of an Airplane Concept of a Stationary High-Altitude Relay Platform (SHARP)", presented at the 32nd Annual Meeting of the Canadian Aeronautics and Space Institute for Aerospace Studies, University of Toronto, Montreal, Canada, May 1985.

6. Glaser, P.E., "The Power Relay Satellite," Plenary Lecture, 44th Congress of the International Astronautical Federation, October 16-22, 1993, Graz, Austria.

7. Criswell, D.R. "Lunar Power System: Research Results and Needs." Proceedings of SPS Conference, August 24-28, 1997, Montreal; Canadian Aeronautics and Space Institute and Societe de Electriciens et Electroniciens de France, Ottawa, Canada. ISBN,0-920203-18-3.

8. Galloway, E.   "The Legal and Regulatory Framework for Solar Power Satellites". In: Glaser, Davidson & Csigi (eds.), *Solar Power Satellites - A Space Energy System for Earth*, 2nd edition. Chichester, England: John Wiley & Sons Ltd, 1998, pp. 425-439.

9. Akiba, R. et al., "International Space Year - Microwave Energy Transmission in Space Rocket Experiment," Institute of Space and Astronautical Science, Report No. 652, September 1993, Kanagawa, Japan.
    Institute of Space and Astronautical Science Task Team Members (47), and NED, Toshiba Corp., Taisei Corp, Nissan Motor Co., Japan Aircraft Mfg. Co., Sharp Corp., Sanyo Electric Co., Hazama Corp., Chubu Electric Power Co., Ozaki Co., and Arianspace, "SPS 2000 Project Concept - A Strawman SPS System.," July 30, 1993.
    Guo-Chin, L., "The Chinese View Concerning Power From Space - Prospects For The 21st Century," IAF-96-R.2.01, Shanghai Institute of Space Power Sources, Shanghai, China, 47th International Astronautical Congress October 7-11, 1996, Beijing, China.

Prisniakov, V.F. et al, "Solar Power Satellite in the Composition of an Orbital Research Complex," Dniepropetrovsk State University, Ukraine, Proceedings of SPS 1997 Conference, August 24-28, Montreal, Canada, ISBN 0-920203-18-3.

10. Bernal, J.D. *The World, the Flesh and the Devil.* Indiana University Press, 1969, p .28.

3

Regional Electricity Development

Stephen R. Connors
Director, Electric Utility Program, M.I.T.

INTRODUCTION

Like many other infrastructure/environment issues, debate on climate change is characterized by (1) complex problems, (2) dispersed solutions, and (3) finite resources.

Quality environmental science obviously plays a central role in identifying reduction targets and describing scientific and economic uncertainties. Similarly, a solid understanding of the energy supply-chain, and its implications for infrastructure development and turnover, is required if cost-effective, coordinated action is to occur. New competitive infrastructure markets -- with their disaggregated technological decision making -- are indicative of the challenges which energy and environmental policy makers face in promoting environmentally responsible and balanced energy markets with finite fiscal resources.

Recent analysis of the New England electric sector shows that aggressive pursuit of energy conservation, renewables, and other forms of non-CO_2- emitting resources will be required to achieve even the most modest carbon dioxide (CO_2) reduction targets. This chapter reviews the technological and institutional challenges of achieving real, long-term reductions in CO_2 and other emissions from the electric sector. Beginning with a New England case study, it addresses factors associated with energy infrastructure turnover, as well as new technology development and deployment. Particular attention is given to policies that promote highly integrated and coordinated, cost-effective, reductions in emissions.

NEW ENGLAND ELECTRIC SECTOR CO_2 REDUCTIONS:
"You Can't Get There From Here"

How difficult will it be to reduce electric sector CO_2 emissions substantially? In 1992, OECD countries agreed voluntarily to reduce greenhouse gas (GHG) emissions to 1990 levels by the year 2000 and presumably keep them there. Current negotiations among participants in the Framework Convention on Climate Change are suggesting reductions of between 10% and 20% below 1990 levels by 2010. Simulations of the New England electric power sector under various technology portfolios indicate that these goals will be technologically -- let alone institutionally -- difficult to achieve.

Exhibit 3.1 shows the basic technological aspects of eight alternative energy mix strategies devised to meet New England's electric service requirements while addressing the issue of NO_x and CO_2 emissions reductions. Each strategy combines the introduction of new natural gas-fired combined-cycle generation with alternate levels of demand-side management (DSM) -- predominantly technology-driven conservation, and for half of the strategies, 1400 MWs of windpower. Each class of technology is phased in during the twenty-year study period (1995-2014). Note that the level of DSM that the region's electric utilities are currently pursuing (hereafter referred t as "Ref. DSM") represents a 10% reduction in twenty-year electricity

Ex. 3.1 Technology Characteristics of Eight Multi-Option
Energy Mix Strategies for the New England Electric Sector

Multiple Option Energy Mix Strategies	New Generation		DSM/Conservation Impacts			
	Nat. Gas	Wind	Peak Gr.	Demand Gr.	Peak Red.	Dem.R ed.
Nat.Gas CC w/Ref.DSM	7140	0	1.06	1.29	-11.1	-9.7
w/Ref.DSM & Wind	7140	1417	1.09	1.32	-11.2	-8.3
w/Double DSM	3540	0	0.460.49	0.57	-21.0	-16.6
w/Double DSM & Wind	3540	1417		0.62	-21.3	-15.2
w/Triple DSM	520	0	-0.08	-0.16	-29.1	-21.2
w/Triple DSM & Wind	520	1417	-0.06	-0.09	-29.4	-19.9
w/Quadruple DSM	0	0	-0.72	-0.56	-37.7	-25.2
w/Quadruple DSM & Wind	0	1417	-0.69	-0.49	-37.8	-24.1
	(MW-2014)		(%/year)		(% from No-DSM Ref.)	

demand, and an 11% reduction in capacity requirements from the no-DSM electricity demand baseline. On the generation side, from 2008 to 2013 roughly half of the region's 6300 MWs of nuclear generation is retired. This means that for the Ref. DSM strategy nearly half of the 7140 Mws of new generation added goes toward replacing the retired nuclear generation, as opposed to meeting growing electricity needs. In the Double DSM strategies, nearly all the new gas-fired generation goes to offset the lost nuclear units. The Triple and Quadruple DSM strategies actually experience a net reduction in total generating capacity by the last year of the study period.[1]

Exhibit 3.2 shows the electricity demand trajectories for each of the four DSM levels. Note how the additional levels of DSM are phased in over time, with the additional efficiency improvements for the Quadruple DSM option ending in 2010. Exhibit 3.3 shows the CO_2, NO_x and SO_2 emissions resulting from these eight strategies (employing four levels of DSM) under moderate fuel cost assumptions. Under the "business as usual" Ref. DSM strategy, by 2014 New England electric sector CO_2 emissions will increase to over 80% above 1990 levels. This sharp increase is caused predominantly by low long-term growth in electricity demand (as indicated in Exhibit 3.1) but is exacerbated by the loss in non-carbon emitting generation sources over time, imported hydropower in 2001, and the retirement of

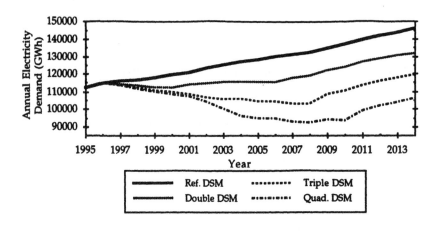

Ex. 3.2 Electricity demand trajectories for four DSM/Conservation options

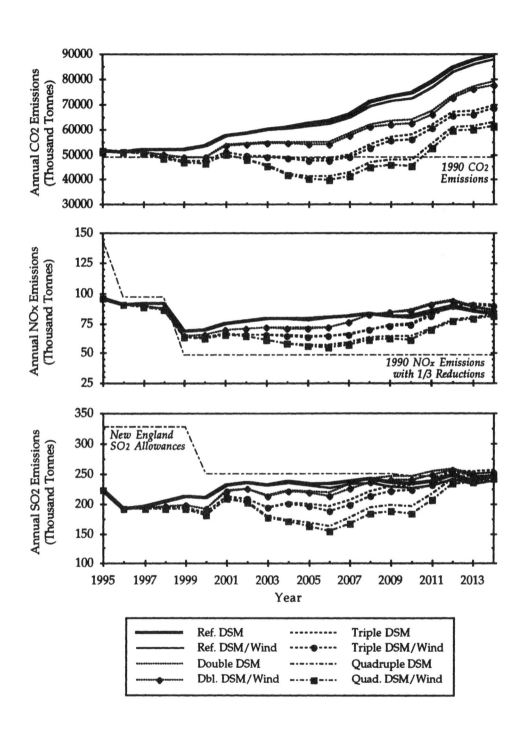

**Ex. 3.3 Carbon dioxide, nitrogen oxides and sulfur dioxide emissions eight
multi-option energy mix strategies for the New England electric sector.**

nuclear units in 2008, 2009, 2012 and 2013. To make up the difference, new and existing fossil generation increases its overall generation, leading to an even greater increase in CO_2 emissions.

High DSM strategies are especially sensitive to the loss of non-carbon- emitting generation sources. While total electricity demand tends to be relatively flat or declining for the aggressive DSM strategies, their CO_2 emissions rise sharply in later years. While Ref. DSM CO_2 emissions are roughly 85% higher than 1990 emissions in 2014, they are still 60%, 40%, and 25%, respectively, over 1990 emissions for the double, triple, and quadruple DSM strategies with windpower. This indicates that reducing CO_2 emissions to or below 1990 levels -- and keeping them there -- will be *very* difficult. Both the aggressive DSM and 1,400 MWs of windpower options constitute technically feasible, but institutionally challenging resource strategies.[2]

For the quadruple DSM strategies, this rise in CO_2 emissions to 25% over 1990 emissions in 2014 occurs rapidly over the final four years. For all the strategies with additional DSM, NO_x and SO_2 reductions in intermediate years are eventually lost, with emissions in some cases exceeding those of the Ref. DSM strategies in the last several years. Why this happens is shown in Exhibits 3.4 and 3.5. Exhibit 3.4 shows the distribution of electric generation by fuel and generation technology for the Ref. DSM strategy without wind. Exhibit 3.5 shows the same range of annual generation sources, but for the quadruple DSM and wind strategy. Note that for the Ref. DSM strategy, when nuclear generation begins to decline in 2008, new natural gas-fired generation is available to replace it. Generation from existing fossil sources, coal and residual oil remains relatively constant. This is not the case in Exhibit 3.5 for the quadruple DSM and wind strategy. Here only a modest amount of new generation is built. Older, less efficient, and more carbon-dense oil-fired generation must be used to make up some of the lost nuclear. These results are symptomatic of an electric industry driven by "capacity mix," as opposed to "energy mix", planning. That is, the decision to build new generating capacity is driven primarily by the need to meet future peak load (MW) requirements, not to attain a balanced generation (GWh) mix. Failure to balance peak load capacity needs with overall fuel mix and environmental performance is one of the first great challenges to a competitive electric market.

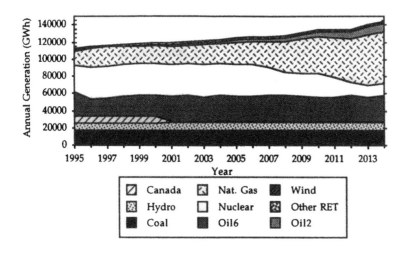

Ex. 3.4 Generation by fuel/technology type for the Ref. DSM and No Windpower strategy[1]

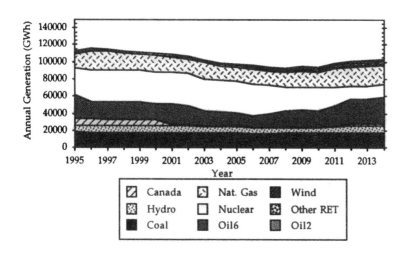

Ex. 3.5 Generation by fuel/technology type for Quadruple DSM and Windpower strategy

In Exhibits 3.4 and 3.5, Canada refers to power purchases from Canada, predominantly hydropower. Oil6 re oil-fired generation. Oil12 refers to distillate oil-fired generation, predominantly peaking units, but also combi during winter. Biomass refers to generation fueled primarily by forestry industry wastes. A significant amoun is currently being burned in existing rankine cycle power plants to meet environmental regulatory requirements.

TECHNOLOGICAL CHALLENGES TO REDUCING CO₂ EMISSIONS

The New England case study identifies the huge challenges industrialized countries' electric sectors will face if significant long-term reductions in CO_2 emissions are required. The fact that it is easier to achieve CO_2 reductions in the electric sector (than in the transportation sector[3]), and that New England is not dominated by coal-fired generation (like so much of the U.S.), indicates that New England might be one of the easier CO_2 reduction case studies. So what is it likely to take, technologically, to meet CO_2 reduction targets in general?

Supply-Side Coordination. The above example shows many of the technological dynamics that need to be considered when looking at how to reduce CO_2 emissions.[4] New efficient gas-fired generation can be considered an interim CO_2 reduction option, but only when it is used to displace existing higher CO_2 emitting units. Therefore, supply-side coordination is key. Natural gas generation did not help New England reduce CO_2 emissions since it either went to meet increased electricity demand, or to replace decommissioned non-CO_2-emitting nuclear generation. New non-CO_2-emitting generation technologies (wind, solar, hydro, biomass, geothermal, nuclear, etc.) therefore must be developed and deployed when meeting increased demand or replacing retiring resources. While many would argue that monopoly utilities historically did a poor job of supply-side coordination (thus leading to competition initiatives), the loss of long-time horizon utility Integrated Resource Planning (IRP) efforts in no way ensures that coordination will be better in the future. Such IRP initiatives not only sought to coordinate near-term resource activities, they also signaled to technology and project developers what level and mix of resources to pursue.

Demand-Side Coordination. The New England case study is also informative, in that is shows us that demand-side efforts must be highly coordinated as well. Not shown above was the fact that aggressive DSM and wind strategies cost more.[5] The present value cost premium these strategies face is due more to the arithmetic of discounting front-loaded capital expenses than to the fact that the technologies cost significantly more. Therefore the timing of resource additions becomes an important aspect of a strategy's cost-effectiveness. Coordination with respect to CO_2 and other emissions reductions occurs on several levels. Similarly, end-use efficiency improvements which displace the least efficient uses of electricity, or can be piggybacked on top of ongoing (new and retrofit) construction, maximize the cost and emissions effectiveness of DSM efforts. As noted above, coordination between supply-side and demand-side initiatives, and among new and existing generation and demand, is also required. When DSM and new generation technologies can be coordinated to displace dirty generation, meet growing demand, and avoid environmental retrofits on aging generation, the benefits of individual technologies can be leveraged. The above case study, as well as previous ones, have shown that failure to coordinate end-use efficiency improvements with supply-side initiatives can lead to increased utilization of older, high emissions generation resources, thereby reducing the effectiveness of DSM and other activities.

Coordination in Space and Time. While coordinating the physical installation of new resources is important from a cash flow standpoint, coordination in space and time is also important when it comes to maximizing the CO_2 reduction potentials of new resources. The real efficiency of a generation resource is a combination of its generation efficiency and the losses experienced in the transmission and distribution of that electricity. In New England, transmission and distribution losses average somewhere around 8% between central station generator and the customer, depending on location, load level, time of year, etc. Therefore an industrial customer generating on-site with an equivalent technology experiences an effective 8% improvement in energy efficiency, not to mention avoided fuel costs, consumption and net emissions. The fact that the industrial customer might choose to co-generate as well further improves overall energy efficiency. DSM and roof-mounted photovoltaics also benefit from this locationally related efficiency increase. Conversely, remote power incurs a penalty by having additional transmission losses. To attain and maintain substantial CO_2 emissions reductions, every little bit will count. Improvements in the efficiency of the transmission and distribution system should therefore also be considered.

Exhibit 3.6 outlines the broad range of technological responses that we will need to take into account as we consider how to coordinate technologies (and their use) in order to reduce CO_2 emissions cost-effectively. The basic architecture represented in Exhibit 3.6 is commonly referred to as the "Distributed Utility Concept." It recognizes that many of the new efficient technologies are small and modular and can be "sited" either at the customer level or at the transmission and distribution system interface. The Distributed Utility concept differs from "distributed generation" in that it recognizes that energy efficiency options, operational protocols along distribution line feeders, and consumer response to real-time spot-pricing initiatives all fall into the distributed architecture as well.

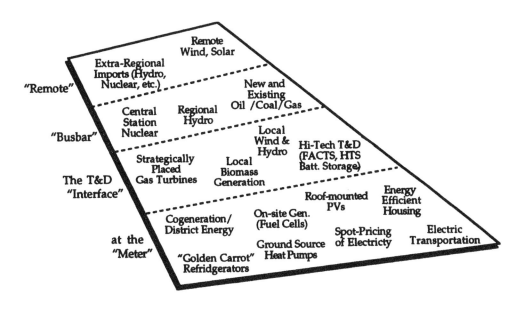

Ex. 3.6 The Distributed Utility concept -- coordinating energy technologies in space and time.

In Exhibit 3.6 technologies are grouped locationally. "Busbar" refers to central station generation, where power flows across transmission lines then to distribution lines to reach the customer. Many utility peaking units are already sited at substations (the "T&D Interface") to support various parts of the grid. DSM and photovoltaics, as well as electrotechnologies and dynamic load-modification efforts take place "at the Meter." On the other end of the spectrum is "Remote" power.

As indicated in the exhibit, renewable energy can fall into several of these categories. Currently, to be cost-effective, wind turbines must be sited in premium wind regimes. In New England many of these sites are in remote areas such as northern Maine. Power flow analysis shows that "remote wind" experiences sizable additional transmission losses (around 20%) if attempts are made to export this power out of state.[6] In contrast, photovoltaics are generally located at the customer site. In New England, the wind resource is skewed toward winter and its associated storm systems. However, New England electricity demand is highest during the summer and concentrated in southern New England. The fact that market prices for electricity will be highest during summer afternoons, and that environmental regulators are looking to restrict summertime NO_x emissions, makes photovoltaics a better "placed" technology in time and space than high-tech wind. An alternative wind turbine technology, which can be cost-effective in lower speed, closer-to-home wind regimes, and which can feed power in at the distribution versus remote transmission system level, begins to look more attractive.

Not in Exhibit 3.6 are other important technological approaches to CO_2 reductions such as sequestration via improved land use and mining practices and power plant CO_2 capture and disposal. These also factor into the technology mix but have fewer coincident benefits. More efficient energy use and generation technologies reduce all the emissions from fossil fuel combustion, not just CO_2. However, given the magnitude of the CO_2 reduction efforts that may be required, these technologies should not be ignored. Cross-national initiatives to facilitate CO_2 reductions, such as joint implementation programs, also need to be evaluated in the light of these multiple benefits. Will exporting efficiency saddle OECD countries with an aging and inefficient fleet of power plants? Will the multiple benefits of accepting costlier but cleaner and more efficient generation technologies make developing countries more likely to participate in joint implementation efforts? Attaining substantial long-term reductions in CO_2 and other greenhouse gases will probably require a complete turnover in the energy capital stock, in generation and end-use alike, including most of the building sector.

These discussions all presume that technology and equipment manufacturers have been sent the right signals such as (1) that replacement technologies have been developed, and (2) that such technologies are readily available for installation and use. As the electric sectors of various nations embrace competition, what structures and signals will be required so that such technological transformations can occur?

INSTITUTIONAL CHALLENGES TO REDUCING CO_2 EMISSIONS

Can competition in the electric sector facilitate both the development and the deployment of these sophisticated and highly coordinated energy mixes into the marketplace? In which direction will competition take the industry and the environment? These are some of the issues discussed briefly in this section.

Quarterly Report Lemmings and Industry Pacesetters. One of the central questions of electric industry restructuring relates to whether it will become an electricity commodity market where every kWh is the same and spot prices rule, or whether it will become a vibrant energy services market where cost, cost stability, reliability, quality, and environmental performance all factor into customers' decisions. Many already criticize corporate boardrooms and financial markets for not taking a long-term view. A commodity market for bulk power would invite the same mentality in the electric sector. A clearinghouse for electricity purchases and sales masks the relative environmental performance of generators, makes efforts to stabilize costs via fuel diversity almost impossible, and virtually ensures that a generic electricity, not electric service, industry will prevail. However, is cheap electricity the ultimate goal? The impetus for competition in energy, telecommunications, and transportation markets has been multi-faceted. High costs, or at least high-cost differentials, have been a primary factor leading to competition in these sectors. But competition has also been seen as a way to invite new capital investment, and new technologies, practices and products into an industry. How can the new competitive market be set up to provide the right rewards to innovators and industry pacesetters?

To invite innovation, the industry must be able to sustain market niches. Product differentiation and customer differentiation must occur. It is generally agreed that wholesale competition alone will not generate the range of market niches illustrated in Exhibit 3.6. Retail competition with customer choice will be much better at creating such niche markets, providing initial markets for advanced energy services, distributed generation, renewable energy, energy storage technologies, and a spate of other applications. More directed efforts at sustaining research and development are being pursued. Renewable and DSM portfolio standards, systems benefits charges, and other such approaches have been promoted and in some cases approved. Customer choice and subsidies for renewables and DSM will play a key enabling role in fostering the development and initial demonstration of innovative technologies. However, several lingering concerns remain.

Promoting Environmental Responsibility. If retail competition is successful in attracting innovative technologies to the industry, will the resulting technological mix be in line with long-term environmental concerns such as climate change? How can their widespread utilization be promoted if and when large CO_2 reductions are required? These are two fundamental questions which have been effectively absent in the electric industry restructuring debate. The first question is difficult to answer with respect to market structure. As monopoly electric utilities disaggregate into separate operating units, they are no longer supporting basic, industry-wide research. Such long-term, strategic R&D efforts will need to be taken up by government. Just as government pursues basic knowledge-related climate change research, it needs to support basic, proof-of-concept solution-oriented research. Industry-wide studies, as well as technology and science initiatives, are important as electric sector "flight simulators" to determine whether we can stay on the cost-effective emissions reduction flight path.

In the event that the electric industry must radically reduce CO_2 and other greenhouse gas emissions, environmental performance constraints on industry will play a central role. Just as competitive manufacturers must adhere to occupational health and safety regulations, a competitive electric industry will need to observe industry-wide environmental standards. "Cap and trade" systems to implement cost-effective emissions reduction strategies will be center stage. The first such system for nationwide SO_2 emissions in the United States has been operating effectively for several years. Along the eastern seaboard, an analogous system for summertime NO_x emissions is in the works. Good science that sets an economically and environmentally effective emissions limit is an important topic with such initiatives. Such performance constraints can have a beneficial impact on industry, promoting a better technological balance than would exist otherwise. With the need to balance multiple emissions markets on top of their normal competitive activities, further development of niche markets is likely to occur as consumers and suppliers seek to reduce their exposure to fluctuating emissions permit prices or avoid the installation of multiple emissions control technologies.

PARTING SHOTS: SLOW AND STEADY WINS THE RACE
We currently do not have the requisite technologies or institutions to implement the CO_2 reductions being talked about in international negotiations by *any* deadline -- let alone 2010. Performance constraints on the competitive market may well accelerate the development and introduction of some of these key technologies, but not by accident. Only with steady, continuous improvements in the efficiency of the energy infrastructure will we truly be able to implement a "no regrets" strategy.

NOTES

The analysis herein was performed by the Analysis Group for Regional Electricity Alternatives at the M.I.T. Energy Laboratory, and was sponsored by the electric utilities, Boston Edison, Central Vermont Public Service, Commonwealth Electric, New England Power Service, Northeast Utilities and United Illuminating, with additional support from the National Renewable Energy Laboratory and the U.S. Department of Energy. Results and their interpretation are the author's and do not necessarily reflect the opinions of the sponsors. Additional assistance from the planning division of the New England Power Pool, NEPLAN was also greatly appreciated.

1. New England Power Pool NEPOOL, Annual Report, 1996, Holyoke, MA.
   In 1996 the six-state New England region consumed 114.7 TWh of electricity. The region's peak electricity demand was 19.5 GWs (in August) and was served by roughly 27 GWs of generation capacity. New England's generation mix is very heterogeneous, with baseload nuclear and coal generation accounting for 26% and 16% of annual electricity output. Intermediate units are predominantly oil and gas-fired steam generation with some new gas-fired combined-cycle units. These comprise 29% of 1996's electricity generation (GWh). Generation from wood, refuse, and hydropower comprised 12% of generation, while peaking units (combustion turbines, diesels, and pumped storage) added another 2%. Imports accounted for the remaining 16% of output. With current electricity demand forecasts, the region's electric utilities do not foresee a need for additional generation stock until around 2005. These expectations are likely to change based upon changes in the economy and energy efficiency standards, the price/demand effect competition in the electric sector might bring, and the future availability of existing power plants such as the region's troubled nuclear units.

2. These high DSM strategies are technologically achievable since it is believed that there are sufficient lighting fixtures, electric motors, and other energy-consuming devices in service that could be replaced with more energy-efficient components. However, with competition imminent, continued support for broad systematic energy conservation efforts is unlikely. Similarly, electric utility efforts to promote large-scale introductions of renewable energy sources, such as windpower, are also less likely. Whether competition will finally help or hinder such technologies remains to be seen.

3. In fact, in transportation CO_2 reduction discussions, electrification of transportation is viewed as a primary option for weaning the industry off fossil fuels.

4. CO_2 emissions are focused on in this chapter instead of greenhouse gas emissions, since CO_2 is the predominant greenhouse gas emitted in the electric sector.

5. See Connors, S.R. and E.T. O'Neill, "No Good Deed Goes Unpunished: The End of IRP and the Role of Market-Based Environmental Regulation in a Restructured Electric Industry", Proceedings of the United States Association for Energy Economics 17th Annual North American Conference, Boston, MA., October 1996.

6. Lacalle-Melero, J., et al., "First Approach to Identifying Wind-Generated Electricity Penetration Limits in New England: Simplified Steady-State Analysis and Economic Evaluation," M.I.T. December, 1994.

Section Two

ENVIRONMENT

4

Saving the Earth - The Whole Earth

John W. Landis
Stone & Webster Engineering Corp., Boston, MA

Asteroids are in the news again. Periodically, members of the press, aided and abetted by authors of science fiction, point out the potentially calamitous effects of one of these celestial bodies striking the surface of the earth. Their definition of "asteroid" is usually broad enough to include fragments as small as a bowling ball. The thrust of their stories generally is that the earth would suffer severe damage from such an impact -- possibly being completely destroyed.

These stories are not overblown. Impacts from debris circling the sun -- even debris as small as a bowling ball -- would frequently have to be characterized as disasters, although most of them would be limited in scope. The impact of a rock the size of the Empire State Building, however, might alter the plant's ability to sustain life.

Since the earth is constantly running into this debris, and since few people have witnessed the effects of these encounters, it is generally assumed that the effects are negligible. Nothing could be further from the truth. Hidden away in the archives of many universities, observatories, public and private laboratories, and industrial companies throughout the world is considerable evidence that the damage is often great.

We know that 65 million years ago a massive asteroid struck the Yucatan region of North America and changed the climate of the entire earth so radically that many animal species, including all dinosaurs, were annihilated. We know that hundreds of substantial celestial fragments have struck the earth since then, culminating in the hit that flattened thousands of square miles of forest in the Upper Tunguska region of diameter, with an energy equivalent of more than 15 megatons of TNT. If it had struck New York City or any other large metropolitan area, more than a million people would have been killed, and millions more subjected to devastating environmental effects.

On the scale of human longevity, the probability of an asteroid disaster is very low, much less than that of being struck by lightening, but it is significant. In view of the potentially horrendous consequences, therefore, taking action to prevent such disasters seems to be a logical step for humankind to take. As far as I know, however, no initiative of this sort has been launched anywhere in the world.

Perhaps humankind is mesmerized by the potential magnitude of this type of disaster. Perhaps we can accept this possibility as our "fate". Neither attitude is sensible. We have the means to set up an effective, although extraordinarily expensive, defense.

We should begin work on it as soon as the requisite international agreements can be developed. The expense is dwarfed by the advantages of its probable success. An asteroid one-half mile in diameter would probably strike the earth with the force of a million Hiroshima bombs. It would be worth a considerable portion of the earth's resources to prevent that.

With our sophisticated solar system exploration, detection and surveillance equipment, we can pick up any substantial asteroid on a collision course with the earth well before (a matter of days) impact would occur. We have guided missiles capable of blasting off into space. We have nuclear and/or conventional warheads with the explosive force necessary to destroy the asteroid -- or deflect it from its course.

An international agency should be established to take sole control of these components and develop them into a global infrastructure which, with the proper procedures, can ensure the safety of the earth. Ironclad agreements limiting the use of this infrastructure to destroying or deflecting space debris or celestial bodies that threaten the earth would be mandatory. Negotiating them would be the most difficult part of this challenge.

Setting up and developing the procedures for this infrastructure would be an enormous endeavor -- one of the largest macro-projects ever undertaken. Like many other macro-projects, it would have great social advantages, including helping to knit humankind closer together. It would demand -- and indeed foster -- the type of worldwide cooperation we need in other arenas to improve the quality of life of the four-fifths of humanity who live in poverty. We have the strongest of all incentives to begin this endeavor: self-preservation. What is lacking is farsighted leadership.

5

Large Climate-Changing Projects

Philip John Pocock
President, Humanitech, Inc.

THREE LARGE CLIMATE-CHANGING PROPOSALS OF THE PAST

Proposal 1: Climate Change by Diverting the Gulf Stream
The Gulf Stream was originally reported in the sixteenth century by European
explorers of the New World, but it was Benjamin Franklin who described it in some
detail. He viewed it as a river within the sea and, with data collected from sea
captains, he constructed a famous chart. During the American War of Independence
Franklin proposed changing the course of the Gulf Stream in order to plunge Great
Britain into a new Ice Age.[1] It was Franklin who first proposed controlling the Gulf
Stream as a weapon of self-aggrandizement.

In 1912 a similar proposal was made by a noted American engineer, Carroll
Livingston Riker. In 1912 he published a book, *Power and Control of the Gulf
Stream*, which contained an ingenious and detailed engineering proposal. His
underlying vision was described in the opening lines of the book's first chapter:

> Climate changes that would double the land value of the Northern
> Hemisphere can be effected by debarring the Labrador current from crossing
> the Grand Bank of Newfoundland, when the Gulf Stream would flow
> practically unimpeded to the Pole, melting every vestige of ice thereabout,
> producing more uniform, warmer winters and cooler summers, and ensuring
> a better climate to every part than is that of New York today. And
> probably, through the melting of the heavy glacial ice cap of the Arctic
> regions, so affect the static balance of the earth in its solar orbit, and the
> inclination of the northern hemisphere, that would produce long twilights
> north of New York, and almost continuous day in Scotland for considerable
> periods, without any period of continued night.[2]

This vision is not surprising for, as Exhibit 5.1 shows, Riker already had bared his promethean nature in the title page of his 1912 book.

POWER AND CONTROL
OF THE
GULF STREAM

HOW IT REGULATES THE CLIMATES,
HEAT AND LIGHT OF THE
WORLD

BY PROTECTING THE WARM NORTH-FLOWING GULF
STREAM FROM THE ONSLAUGHTS OF THE ICE-COLD
SOUTH-FLOWING LABRADOR CURRENT

MAN CAN CONTROL ALL

CAUSE AND EFFECTS OF OCEAN CURRENTS
THE EQUATORIAL AND THE POLAR FORCES

BY
CARROLL LIVINGSTON RIKER, M.E.
BROOKLYN, NEW YORK
1912

FOR SALE BY
THE BAKER & TAYLOR CO.
TRADE SELLING AGENTS
3437 E. 17TH STREET, NEW YORK

Ex. 5.1 The title page of Riker's book.

"Man Can Control All" was a catchy modernist sentiment, but the man to whom Riker dedicated his book was, at the time, the world's major regal power-holder and the man in whose empire Newfoundland was ensconced (Exhibit 5.2).

THIS WORK IS MOST RESPECTFULLY DEDICATED TO HIM WHO NOW

HOLDS THE POWER TO PHYSICALLY CONTROL THE WORLD.

ITS CLIMATES, ITS HEAT, AND ITS LIGHT.

His Royal Highness and Imperial Majesty

King George V

Ex. 5.2 Riker's dedication of his book had a Renaissance air about it.

Riker was a doer as well as a dreamer. Although his imagination could soar, he could engineer it back to earth. As a youth he was fascinated by the sea, its currents and it utility. At the age of seventeen he designed the hull of the steamboat *Charleston*. At twenty-one he designed the first refrigerated warehouse in New York. He designed the most powerful dredge of his time. In 1907 his experience was enlisted in the building of the Panama Canal. During the Spanish-American war he designed a new type of torpedo. In 1972, Gaskell writes:

> Riker ... lived a life full of new ideas and inventions, mainly concerned with the sea.... A lifetime of experience in large-scale earth moving, and in matters pertaining to the oceans ... gave Riker the idea for what turned out to be a nonstarter, but which might have been his greatest civil engineering project of all. This was the scheme to control the Gulf Stream, and it was no vague idea, but rather a detailed plan backed by well thought-out methods of attaining the desired objective, and by proposed experiments to find necessary information.[3]

The cold Labrador Current from the north intermingled with the warm waters of the Gulf Stream, and the goal of Riker's proposal was to build a 200-mile-long jetty from the eastern tip of Newfoundland out along the Grand Banks. The purpose was to inhibit the intermingling of the two currents. The foreseen benefits were three:

- the elimination of fogs in the Newfoundland area;
- the removal of icebergs from shipping lanes;
- improved climatic conditions in the countries near the Arctic Circle.

Exhibit 5.3 illustrates Riker's vision of a new system of currents in the North Atlantic. Riker believed that preventing contact between the two currents would cause the Gulf Stream to continue to flow on its eastward course and then it would flow northward and divide at the southern tip of Greenland, thus providing two warm streams on either side of Greenland.[4] Gaskell comments that the warm current along the west side of Greenland might melt a large part of the Greenland icecap, and it is possible that Riker's project might have produced a much milder climate in Scandinavia and Siberia. In view of Riker's ingenious ideas for constructing the jetty, Gaskell called this "a magnificent project" but noted that there was probably not enough information on hand to ensure its success -- something that Riker realized.

Riker proposed that before attempts were made to build his project, careful preliminary studies must be made of the currents to assess the practicability of the project and the climatic results if the project was achieved. Thus Riker proposed a feasibility study and the journal *Scientific American* offered to furnish a corps of experts to accompany Riker on his investigations of the Labrador Current. Several companies offered support, and Gaskell thus concluded that the project had much "well-informed support". But was there well-informed support for Riker's title declaration: *MAN CAN CONTROL ALL*? The answer has to be *Yes*!

Riker was born in 1853, just the right time to imbibe all the high optimism inspired by the revolution of Modernism, optimism resulting from the industrial and scientific revolutions and the accumulating wealth of the time. As Europeans approached the year 1900 they sensed the coming glory of the millennium as they experienced what they believed to be its harbingers: the automobile, the airplane, electric power and lighting, new materials, the telephone, huge structures. The French celebrated the *fin de siecle* by building the Eiffel Tower in 1889, and by celebrating the anniversary of the French Revolution.

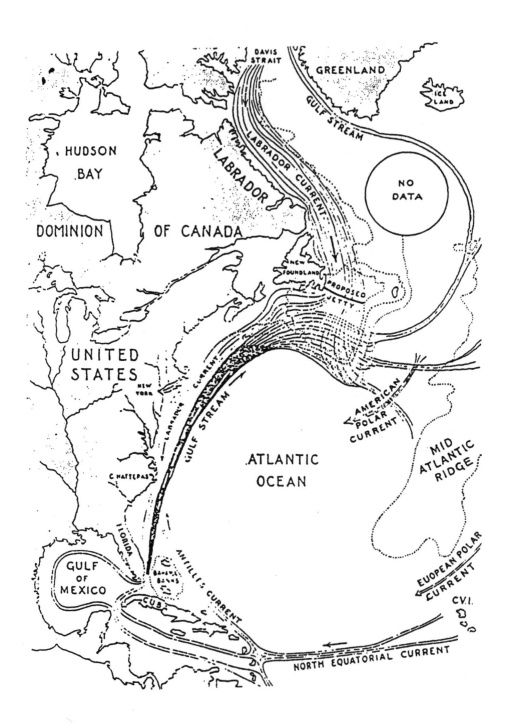

Ex. 5.3 Sketch showing Riker's vision of a changed Gulf Stream course

The sense that a new modern world could be built as an exciting and proper preparation for the next millennium colored the activities of Western Man. The Australian art critic, Robert Hughes, has given us a compelling account of how this revolution transformed art and architecture, just as it transformed people's expectations and the conventional wisdom of the times.[5] There was widespread optimism, a time of growing progress had begun, great times lay ahead. Carroll Livingston Riker was a true child of his time. He would have been thrilled by the triumphalism of the 1893 Chicago World's Fair where, in The Palace of Mechanical Arts, one could see "a forest of moving mechanisms". The illustrated book about the fair stated: "Machines are the paramount symbol of modern material progress. The display of them at this World's Fair is the greatest ever seen on earth."[6] It was not only faith that could move mountains.

While the successful accomplishment of the goal of Riker's project would be beneficial to some countries, it would be considered an act of war by others, and such a view was expressed in a German novel published in 1923. Gaskell tells of a German author, J.E. Kiesel, whose fictional work described a seven-year war waged by the U.S. against Europe.[7] The German hero of this confusing work devised a plan to divert the Gulf Stream, by means similar to Riker's, and sold the information to the Americans. Gaskell claims that Kiesel appears to have hated England and, like Ben Franklin, wished to deflect the Gulf Stream to freeze Britain. Some commentators suggested that Riker, fearful of European preparations for World War I, believed that his project might prevent such a war by freezing the Kaiser's land.

In January 1913 Riker presented a *Conspectus of Power and Control of the Gulf Stream* to the President and Congress of the United States and "for the officials of Interested Foreign Countries and others."[8] Although there was some informed support for his project, World War I was about to begin and resources were directed elsewhere.

PROPOSAL 2: A SOVIET CLIMAGE-CHANGING PROJECT: DAMMING THE BERING STRAIT

Whereas Carroll Riker, heady with the turn-of-the-century optimism of modernism and with his own engineering success, declared on the title page of his book, *MAN CAN CONTROL ALL*, the title of a 1973 book by a Soviet engineer, Pyotr Borisov, is titled, *Can Man Change the Climate?*[9] Yet the book opens with the obligatory all-in-caps quotation from Lenin:

...UNTIL WE KNOW A LAW OF NATURE, IT, EXISTING AND ACTING INDEPENDENTLY OF AND OUTSIDE OUR MIND, MAKES US SLAVES OF "BLIND NECESSITY". BUT ONCE WE COME TO KNOW THIS LAW, WHICH ACTS (AS MARX REPEATED A THOUSAND TIMES) INDEPENDENTLY OF OUR WILL AND OUR MIND, WE BECOME THE MASTERS OF NATURE.

V. I. LENIN

Borisov first floated his idea in 1957,[10] but by 1973 experience had shown Soviet citizens that "mastering nature" was only a slogan arising from Lenin's revolutionary hubris. The designer of Borisov's dust jacket hinted at this by putting

the title's question in lowercase letters.[11]   The writer of the book's foreword, the
geographer Dr. S. Y. Geller, also showed a restraint unknown to Lenin; he wrote:

> .... only since the end of the last century, when the enormous importance of
> the Arctic in forming the climates of the Earth became apparent, have
> researchers turned their thoughts towards elaborating methods of reducing
> glaciation and then of completely eliminating the ice of the Arctic Ocean in
> order primarily to change the climate in the high and middle latitudes of the
> Northern Hemisphere. .....   P. M. Borisov, a Soviet engineer ... has
> devised a very interesting project for changing the climate on a planetary
> scale.

Dr. Geller then describes Borisov's proposal:

> [Borisov recommends] liquidating the ice sheet of the Arctic Basin.   The
> intention of Borisov's project is to create a direct flow of warm Atlantic
> waters through the Arctic Basin and at the same time prevent the flow of
> cold waters of this basin into the Atlantic.   This, he suggests, is to be
> brought about by transferring the surface Arctic waters to the Pacific by
> means of powerful pumps installed in the Bering Strait.   The removal of
> cold waters and the inflow of additional warm Atlantic waters should
> prevent the formation of ice in the Arctic Basin, and thus create a warmer
> climate.

Geller then expresses professional caution:

> However, a good deal of [Borisov's project] is still debatable and obscure.
> The reaction of Nature over vast territories to the change in glaciation of the
> Arctic Basin cannot as yet be determined, nor can this reaction be taken for
> granted....   The many questions that must be answered to solve so
> tremendous a problem as the planetary melioration of the climate can
> scarcely be enumerated.   However, there are reasons to assert that this
> problem is acquiring very real and tangible substance... The author's
> research has made a great contribution to the advancement of this question.

By page 111, Borisov, in a chapter titled *A Gulf Stream in the Arctic Basin,*
considers some criticisms of his project:

> And so we have attempted to show that a Polar Gulf Stream project is a
> feasible undertaking. However, some scientists contend that [the project is
> flawed]....   In essence, all these arguments boil down to this: It is
> impossible to create a Polar Gulf Stream, but even if it were possible, it
> would not be expedient. We need quantitative data to refute all these
> arguments. Let us examine them in greater detail. [He does so.]

Finally, after a chapter titled "A Gulf Stream in the Pacific Ocean", he moves on
to predict favorable changes in the weather in North America, particularly in
Canada[12] -- and finally he gets down to a detailed description of his damming of the

Bering Strait, in a chapter entitled "Practical Implementation Problems". Gaskell comments that, as in all grandiose projects, the difficulties are camouflaged by lots of detail, such as intricate drawings of Borisov's dam and its pumping system.[13]

Borisov concludes his book thus:

> Any large-scale climatic amelioration cannot be confined within the borders of any one state. Hence, the problem of the major climatic changes inevitably overrides national boundaries and occupies a place among the urgent international social and political problems.... The amelioration of the Arctic will undoubtedly open a new chapter in international relations. It concerns not only the USSR, Canada, Greenland, Denmark, USA (Alaska), whose territories are situated in the planers coldest regions, but also many countries in tropical and desert zones.

Gaskell comments in his 1972 book that it would be possible for Borisov's dam to be built, and he adds that with the rapid growth of population it might well be advisable, one day, to open up the vast areas of land bordering the Arctic Circle. "If the fierceness of the weather could be eased by modifying ocean circulation, the long dark winter nights might be bearable."[14]

A parting thought: What would the situation be if: (1) Russia had not sold Alaska to the U.S.; (2) Pyotr Borisov had the same relationship with Stalin as did Trofim Denisovich Lysenko (1898-1976); (3) Stalin decided to construct the dam; (4) there had been no way of stopping him. The world's politicians did not stop Japan from invading China, nor did they stop Mussolini from his African conquests. Why would they have stopped Stalin from building this "exciting" dam?

In the future, will the world be faced with a similar situation? There is no way to give a definitive answer to the latter question, given the corruptive effects of power, the shifting locations of power, and the demonstrated reluctance to restrain rogue leaders. The possibility of such a situation places a high priority upon developing the ability to predict successfully the effects of such projects.

PROPOSAL 3: A SUGGESTION FROM JAPAN: DAMMING THE DRAKE PASSAGE

In 1970, Keiji Higuchi, a Japanese professor at Nagoya University's Water Research Laboratory, presented a paper, *A Possibility of Constructing a Dam to Change the General Oceanic Circulation.*[15] Higuchi was encouraged to suggest such a project for two reasons. First, he was impressed by progress in the research field of numerical experiments on general circulation models of the atmosphere and oceans, citing work by Takano[16] and a paper by Manabe and Bryan[17] as examples of such progress. Second, Higuchi took courage from Borisov's work and declared that it was possible to change the climate because he believed it possible to calculate the effects of large climate-changing projects. He took further encouragement from the claim that the climatic effect of Borisov's goal of ridding the Arctic of ice could be calculated, as shown by work done by Arakawa.[18] One can calculate the effects of interrupting ocean currents; thus one should be able to calculate if such projects were beneficial.

Higuchi declared that "one of the important aspects of future research" was to propose a "model to change world climate in order to perform numerical experiments". In his talk he proposed the following project for study by modelers: Building a dam made of icebergs across the Drake Passage separating South America from Antarctica. He believed that this would be an interesting project for study because it would block the westward passage of the Antarctic Circumpolar Current, a current flowing through the Pacific, Atlantic, and Indian oceans (see Exhibit 5.4). He believed that this dam would "have the effect of changing the ocean circulation on the global scale." But he made no speculation as to what the changes would be nor who would benefit by this project; that would be left for model studies. He added, however, "it can be said that the Drake Passage is a key point to changing the general circulation of the ocean and the atmosphere." He discusses in detail how the dam would be built with icebergs and he ends with these concluding remarks:

It would be quite an interesting problem to perform numerical experiments on the general oceans and atmospheric circulations under conditions of closing the Drake Passage. If the result of numerical experiments shows that construction of the dam produces a better climate condition over the world, this process can be considered as a good method for artificial control of the climate. If it shows that the construction of the dam produces such a cold climate as a glacial period, this process can be considered as one of the causes of the Ice Age.

Takano and Higuchi made an initial calculation of the effect of this dam on the volume of transport alone, and this later addendum to Higuchi's paper ended with the sentence: "It is hoped that much more extensive calculations will be done with success in the very near future so as to be useful in planning artificial control of the climate." No study was made to uncover more recent Japanese climate-changing projects.

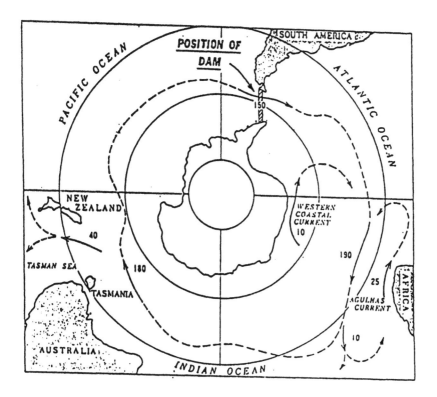

**Ex. 5.4 Position of proposed ice dam for impeding the
Arctic Circumpolar Current.**

THE "DREAMS OF ENGINEERS": THEIR IMPORTANCE

The dreams of engineers, like the hypotheses of scientists, are the seeds from which some great things arise. Such insights, to quote Canadian philosopher, Bernard Lonergan, are "a dime a dozen" but without them we would be reduced to sterile data gatherers. These dreams and hypotheses, like the charcoal sketches of painters, are like the sudden idea for a novel that came to Henry James as the result of a chance remark at a dinner party. For example, as Higuchi suggests, the dreams of engineers for climate changing should be studied in order to develop modelling skills, to gain understanding of our ignorance and our lack of scientific knowledge, and to open the portals of our minds to serendipity. Careful study and analysis of climate-changing projects, such as the above three, would be of great heuristic benefit and a necessary complement to the pursuit of scientific understanding of the relevant processes.

After World War II, one of the founders of the German Rocket Society, Willy Ley, wrote a book, *Engineers' Dreams*.[19] The "dream" in the first chapter of Ley's book has already been "engineered down to earth": it is called the Chunnel, a realizable project that Frank Davidson was instrumental in bringing about by sharing this dream and helping it become a physical reality. But let us look at some other dreams.

There are many macro-engineering proposals out there that might attract the attention of a powerholder with the means of project initiation. For example:

♦ Damming the Strait of Gibraltar. This is just one element of a project called Atlantopa, an ambitious project for changing the Mediterranean.[20] The result: huge power development and a large area of land reclaimed from this sea. This plan was described in a 1928 paper by Herman Sörgel, an architect employed at the time by the Bavarian government. Will attempts be made to sell such a project to the EU governing body?[21]

♦ In 1935 Sörgel conceived another huge macro-engineering project: the creation of a central lake in Africa.[22] At this time details of the project had been described in German engineering and political journals. There was a major difficulty: the project would be impossible to carry out unless all of Africa was one political unit. During World War II, when Germany and the Soviet Union were not yet at war, the German Foreign Minister, Joachim von Ribbentrop, approached his Soviet counterpart with what Ley calls a "strange document". It described "spheres of influence" after the war, and the Germans stated that the Russians could have control of Asia if Germany could have control of Central Africa, leaving some parts for Italian control. The Russians did not sign the document and the Germans did not press them to do so.

♦ About 1925 a Frenchman, Pierre Gandrillon, embarked on his "Plan for Palestine", a major power development project for the Valley of the Jordan.[23] Ley comments: "The political future of Palestine was highly uncertain during the period from 1925 to 1935 and nothing was done about Gandrillon's plan." In 1943-44, an American soil expert, Dr. Walter Clay Lowdermilk, went to Palestine and, without knowledge of Gandrillon's project, came up with an equivalent plan for what he termed the Jordan Valley Authority, or JVA.

♦ Gaskell describes a "bizarre" scheme for the Gulf Stream, one that would make for a milder climate in the north, for Britain and Scandinavia.[24] The idea was to put turbines to work in the Gulf Stream in the Florida Straits and generate power. [Editors' note: See Chapter 1]

Such projects will always be proposed as long as human imagination is allowed to be exercised, but they are portals of discovery and serendipity and should be treated so. Dreams, visions, insights, or hypotheses though they may be, we should attempt to convert them to dreams constrained by reason. It was the Spanish painter Goya, who proclaimed: "Imagination deserted by reason creates monstrosities. United with reason, imagination gives birth to great marvels and true art." René Dubos was inspired by this observation to title one of his books *The Dreams of Reason; Science and Utopia*, a book analyzing the relationship between scientists and the lay public.[25]

THREE REVOLUTIONS AND THE DISCOVERY OF THE CENTRAL PHYSICAL PROBLEM OF OUR TIME?

A Revolution of Consciousness: Seeing Earth for the First Time
A Canadian, Marshall McLuhan, predicted that once man gained sight of earth from outer space and brought back a photograph, earth would become an art object, a source of aesthetic pleasure. This new image would inspire more delight than the

sight of our moon, for people would finally see the green-white-blue beauty of our unique homeland; awesome beauty because, surrounded by the blackness of surrounding space, isolated in the uninhabited Milky Way galaxy of at least 100 billion stars, it alone possesses life and its remembered joys. McLuhan's insight was correct, for Alan Shepard, the first of the American astronauts to view the earth in the surrounding blackness of space, said, "What a beautiful view!"[26] The astronauts became poetic in their comments as they viewed earth. Soviet cosmonauts agreed with one of their colleagues who said "Our planet is uncommonly beautiful and looks wonderful from cosmic heights." They saw Mother Earth from afar and were awed by her beauty -- an image that has become an icon for our times.

On February 20, 1962, John Glenn became the first American to orbit earth and to photograph it. The captured view was arresting and *Time Magazine* replaced the "Man of the Year" with a cover photograph of earth -- the "Planet of the Year". Today, the cover of the November 1996 issue of *National Geographic* contains a photograph of earth, indicating the issue's cover story: "The Astronauts' View of Home". It is estimated that astronauts have taken some 300,000 photographs of the earth -- a beautiful sight for us earthbound creatures to behold. It shows us where we are; finally an abstract understanding is replaced by sensory experience.

The Copernican Revolution, the victory of a simplification wrought by mathematics, delivered a shock to the consciousness of human kind: we were not enthroned at the center of creation. Common sense had it that the sun rose and fell, thus circling our days with its light and warmth. It was comforting to our ego to believe the heavens moved around our unique species as it delivered delight and succor. But the truth was that our small unique planet revolved around our sun,[27] a truth more startling than the roundness of earth. Both truths are contrary to common sense, which suggests we are not riding in a rotating home on a cosmic ferris wheel.

"Revolution", meaning motion in a circular path or a large and significant change, is a much used word, because we inhabit a world of continuous change. As well as the Copernican Revolution, we have had the Cartesian Revolution, the Scientific Revolution, the Industrial Revolution, the Revolution of Rising Expectations and many more -- and many more to be. It is not surprising, therefore, to find "revolution" a not uncommon word in book titles. For example, a search of a database of books in print results in a count of the number of book titles with the following words:[28]

- Revolution 8,047
- Mind 6,096
- Brain 3,303

"Revolution" is popular, but not as popular as "love" or "lov" (e.g., loving):
- lov 26,174
- love 19,892

One is reminded of Dante, who was a strong advocate for the use of reason and was fascinated by astronomy, yet, as a poet, wrote of "the love that moves the sun and other stars".

THE SECOND REVOLUTION: GETTING TO KNOW AND TO LOVE MOTHER EARTH

Until our time people had some abstract knowledge that the earth was round, but this knowledge could be ignored as it had operational utility for few. The majority of people could operate as if the world was flat, and for all practical purposes, extended to infinity. Now everyone is conscious of it being round and beautiful (as witnessed, for example, by the passion for travel). In our minds, the icon image of earth is an object of aesthetic attraction. Its presence has drawn not only the attention of EveryPerson, but of the scientific community whose widespread and rapidly growing attention to our global system has led to a second major change, *a second revolution*: a revolution arising from the growing international search for scientific understanding of earth, of nature's technology, whether of sea, storm, sky or of earthbound life itself.[29] As the ancients pointed out, attraction leads to attention, attention leads to the pursuit of knowledge, and increasing knowledge can lead to love. Knowledge always precedes a committed love, thus the expansion of our scientific understanding is inducing in us a loving regard for our small galactic home. Growing numbers of people love earth, want to travel around it, want to see it bettered, want to see it survive.

This second revolution, like all revolutions, gave birth to unforeseen results, of movements, of goal-directed activities. This revolution gave birth to the environmental movement and sustained the rapid growth of its concern. It stimulated a desire among the world's scientists to expand our understanding of our global system. This revolution resulted in a growing love and concern for Mother Earth and its biosphere of living beings. A new familiarity was abroad. Our planet became known as Spaceship Earth. We realized that McLuhan was right: we are a global village. The rubric *Act Locally, Think Globally* was accepted by those with a consciousness of the beauty, worth, and village-like nature of our small spherical home. We no longer lived on a flat earth of almost immeasurable extent but, helped by transportation and information technology, we became more conscious of being neighbors, globally. More and more, flat-earth engineering and policy making would become increasingly irrelevant as the problems to be faced became more globally interactive in extent.

But there has been a great and troubling surprise, for scientists, in their diligent study of the global system have uncovered physical evidence suggesting future global catastrophes: an increasing global warming. If this trend continues social and economic chaos might overtake us. When the tabloid press catches up on this issue, there will be headlines at our check-out counters such as: *Mother Nature's Revenge-The Death of Nature as We Know It.*[30]

THE THIRD REVOLUTION: THE REVOLUTION OF NATURE'S RESPONSE

The mindset of the Renaissance led to the Revolution of the Renaissance. As Lisa Jardine has superbly documented with fascinating detail, the mindset of the Renaissance inspired a lust for power and wealth by means of trade and conquest; arms were the instruments of policy making, a harbinger of Mao's definition of politics.[31] In addition, the possession of wealth by the bankers and merchants -- and the love of ostentatious display by the powerful -- led to the support of great art, architecture, music, and scholarship. The view of man as master of this earth was comfortably enlarged, and "worldly goods" became an index of worth. The Industrial Revolution was an inevitability.

The success and global expanse of the Industrial Revolution, supported by a growing interaction with the insights of the Scientific Revolution, have faced us with a threat unique in history. In past years, the great and notable extinctions of human lives and habitats have been due to calculated military actions or genocidal passions. This has always been a source of our great fears, such as we experienced during the recent Cold War: the fear of nuclear extinction or nuclear "winter". Although "world wars" have recently been waged, their physical effects were more macro than truly global. But today, although there are haunting threats of terrorists developing and using weapons of mass destruction, we are faced with a historically unique global threat: the feedback response of nature transgressed by global warming (i.e., from greenhouse gases) resulting from our "successful" human activities. The haunting fear is that a severe climate change will occur that will have devastating social and economic effects around the globe, revolutionary effect -- a situation of comfort only to the current proponents of Armageddon or the growing crowd of millennial cosmic cranks.[32] The bearers of the sign, *THE END IS NEAR*, may now pose as the heralds of a new Ice Age, for some believe that an ice age could be a result of our human success. And if that was not worry enough, Allan Stone, the Professor of Modern History at Oxford University, predicts that we are now in a New Dark Age, a view shared by the popular French historian, Alain Minc.

At the present time we are witnessing what is probably the largest scientific and policy research exploration of all time, and this is in response to the threats posed by global warming, by climate change, and their inevitable consequences. This activity is supported by governments around the globe and is gathering a widening support from all professional quarters. Recently a senior U.S. government official was asked whether or not this sudden large sprouting of intense and widespread activity was just a bandwagon activity led by self- interested scientists and policy advisors, or was there, behind it all, a deep and serious government concern. He answered that the latter was the case. Governments are concerned. They are aware and puzzled by the revolution of nature's response to the impact of our stewardship. If nature's response to our actions will lead to a climate change wrought by an increasing warming of our atmosphere, as experts believe, then this is the fact: humans may be facing their greatest physical problem as they enter the next millennium.

In place of a lengthy review of the extensive global warming and climate change literature, the following is a short list of printed information and World Wide Web site addresses that should lead to an extensive understanding of the extent and seriousness of the revolution of nature's response.

THE CENTRAL PHYSICAL PROBLEM OF THE 21st CENTURY? GLOBAL CLIMATE CHANGE

If global warming increases to the degree that some predict, and if this warming leads to the predicted catastrophic physical changes of the global physical system, then climate change may be called the central physical problem of our time. It is not the purpose of this chapter to present evidence to remove or maintain the word "if" in the above. The purpose of this section is twofold: (1) to establish the fact that there is an intense and growing investigative activity regarding global warming and climate change; (2) to present a number of current information sources that should assist people in coming to their own answer to the above question. The most remarkable source of information is the World Wide Web, and Web addresses are given for a number of sites that collectively present a portal through which a more comprehensive view may be gained than could be presented here.

This position is taken because experts agree that the first step in the process of problem-solving decision making is the gathering of intelligence, that is, gathering information that is within a given context. What is frequently called "information" is really "intelligence", as it is in industrial espionage or "spying".

The primacy of relevant information was noted by MIT's Jay W. Forrester, who described the role of decision making as follows: "Management is the process of converting information into action. The conversion process we call decision making."[33] The management scientist, Herbert A. Simon, described the decision-making process as a three-step process of *intelligence, design, choice.*[34] In our case *intelligence* (that is, information within an appropriate context) would be gathered about the problem of climate change; the *design* process involves searching for means that might solve the problem, for example, by reducing carbon dioxide in the atmosphere; *choice* involves deciding upon the criteria for choice, if choices are available -- otherwise ones must be developed. Simon sums up this decision-making structure as follows:

> [I distinguish] three phases of decision making: intelligence, design, and choice -- processes for scanning the environment to see what matters require decision, processes for developing and examining possible courses of action, and processes for choosing among courses of action.[35]

Simon's insight helped stimulate the birth of computer-based support for decision making.[36]

The first step in decision making is acquiring relevant information (i.e., intelligence) about the problem. Here there must eventually be a division of labor between the scientist and the engineer, for the role of the engineer is normative and that of the scientist descriptive. Simon, in his 1968 Taylor Compton Lecture, *The Sciences of the Artificial,*[37] defined the engineer as follows:

> The engineer, and more generally the designer, is concerned with how things ought to be -- *ought* to be, that is, in order to *attain goals*, and to *function....* Engineering, medicine, business, architecture, and painting are concerned not with the necessary but with the contingent -- not with how things are but with how they might be -- in short, with design.... Everyone designs who devises courses of action aimed at changing existing situations into preferred ones.... Design, so construed, is the core of all professional training: it is the principal mark that distinguishes the professions from the sciences.

The engineer's goal is to help establish or restore a norm within the constraints of the known laws of nature, the laws of the society, the available resources, and his or her personal experience, knowledge, know-how and problem-solving abilities. A problem that "ought not to be" exists when people *need* something and do not know how to obtain it.[38] It may be the entrepreneur who perceives an unmet need as a well-disguised opportunity, but it is the engineer who is essential[39] to the design the overall solution. People need food, shelter, transportation *et al*, and the engineer always plays an essential role in the material aspects of designing, producing, and supplying a solution. Often it is the dream of an engineer that leads to a solution. But today the engineer is inevitably working in relationship with the scientist.

Simon also described the role of science:

> A natural science is a body of knowledge about some class of things -- objects or phenomena -- in the world: about the characteristics and properties that they have; about how they behave and interact with each other.... The central task of a natural science is to make the wonderful commonplace: to show that complexity, correctly viewed, is only a mask for simplicity; to find pattern hidden in apparent chaos.... This is the task of the natural sciences: to show that the wonderful is not incomprehensible, to show how it can be comprehended -- but not to destroy wonder.

> With goals and "oughts," we also introduce into the picture the dichotomy between normative and descriptive. Natural science has found a way to exclude the normative and to concern itself solely with how things are. Can or should we maintain this exclusion when we move from natural to artificial phenomena, from analysis to synthesis?

Simon's query is in response to the proclivity of many members of distinct groups, be they intellectuals or athletes, to seek to maintain a tribal-like separatism. Nevertheless the information needs of two distinct groups -- the "scientists" and the "engineers" -- are not totally different.

Below are references to information sources regarding global change, such as global warming and the policy issues involved. The references cited are intended as a springboard for entry into the veritable ocean of information that lies awaiting.

For those in the engineering profession, Gregg Marland discusses an important question of today in a paper: *Could We/Should We Engineer the Earth's Climate?*[40] This paper is a very good and blessedly brief introduction to the central question for macro-engineers. Marland's paper is published together with several other papers, ones previously presented at a 1994 symposium organized for the 1994 annual meeting of the American Association for the Advancement of Science (San Francisco, February 19 and 20, 1994). At this conference, ten thoughtful experts addressed the issue of whether or not "global engineering" or "geoengineering"[41] should be the means of attack on the climate change problem. Several issues were considered such as "could we" and "should we". Attempts were made to distinguish clearly between the technical and social issues. Marland's paper from this AAAS conference and that of four others have been published in the journal *Climatic Change* for July 1996; the table of contents of this important journal is on the Net.[42]

Marland states that there was no consensus reached at the above AAAS meeting and he concludes his paper as follows:

> [The papers] form part of an important dialogue. They provide analysis not advocacy. They raise the technical possibilities while emphasizing the risks and posing the ethical concerns. They make it abundantly clear that relying on these approaches now would be irresponsible, given our current state of understanding of the climate system. And yet, we have to be concerned whether society has the will to confront, at their source, our environmental impacts in such a way as to avoid serious long-term consequences.

For engineers, a good supplement for the papers in the above issue of *Climatic Change* is the 1992 massive report *Policy Implications of Greenhouse Warming: Mitigation, Adaption, and the Science Base,*[43] published by the National Academy Press. Those involved with policy should also include a 1991 report produced by the panel producing the latter report.[44]

To perceive the monumental efforts being devoted to pursuing the problem of global warming and other global changes, an individual is well-served by entering the World Wide Web to surf, for thanks to hypertext- a dream of MIT's dean of engineering, Vannevar Bush -- serendipity becomes a fruitful guide. But the majority of the material is focused on scientific understanding and policy implications and choices, which is as it should be, for understanding and thought should precede action. Beware, for there is Global NewSpeak used here with abandon. Some suggested stops are offered in the following table.

U.S. Global Change Research Program -- USGCRP
http://www.usgcrp.gov/

For information about U.S. activities, this is a monumental Web site. It has links with the principal U.S. sites, and there are important links to international sites. For example, access is given to The Intergovernmental Panel on Climate Change (IPCC); its 1995 reports, from the three working groups, are a mouse click away. The extent of USGCRP's influence is witnessed by the fact that it has sponsored 361 research institutes, mainly at universities. There is much more.

Global Warming International Center-GWIC
http://www2.msstate.edu/~krreddy/glowar/glowar.html

At this site is the statement: "GWIC is the international body disseminating information on global warming science and policy, serving both governmental, non-governmental organizations, and industries in more than 120 countries. It sponsors unbiased research supporting the understanding of global warming and its mitigation.... GWIC sponsors the annual Global Warming International Conference & Expo, and the Executive Workshop on Industry Technology and Greenhouse Gas Emission, which facilitates international exchange and provides the most up-to-date hands-on workshop for corporation and utilities executives." This Web site contains the *World Resource Review* (WRR), "an international refereed journal devoted wholly to scientific and policy studies of Global Warming, World Resource Management, and Global Change." This site also contains the *Global Warming Revue*, a huge bibliography of the associated international journal - and more.

The U.S. Global Change Research Information Office- GCRIO
http://gcrio.ciesin.org/

GCRIO "provides access to data and information on global change research, adaptaion/mitigation strategies and technologies, and global change related educational resources on behalf of the US Global Change Research Program (USGCRP) and its participating federal agencies and organizations. GCRIO is implemented by The Consortium for International Earth Science Information Network (CIESIN)." There are many hypertext links here and every subject is covered; for example, Our Changing Planet, the FY 1996 Global Research Program document, a supplement to the President's Fiscal Year 1996 Budget. What becomes immediately apparent is that the global change cadre have been forced, due to its rapid growth, to invent and use a NewSpeak; thus to facilitate the tyro, this site includes a list called, *Global Change Acronyms and Abbreviations-* 361 of them! For example, no longer do ARM, EROS, GEMS, GOALS, MAN, PAGES, SCOPE, TAO, WEAVE... mean what they do in NormalSpeak. And what about: AFGWC, CZCS, EO-ICWG, GNMDF, IWGDMGC, or UNFCCC? GCRIO has the civility to make much of the list "clickable".

Atmosphere Radiation Measurement (ARM) Program
http://www.arm.gov/docs/about.html

This site presents another list of acronyms and abbreviations (not clickable) and leads one to learn about the ARM Program, which is "a multi-laboratory, interagency program that was created in 1989 with funding from the DOE. The ARM program is part of DOE's effort to resolve scientific uncertainties about global climate change with a specific focus on improving the performance of general circulation models (GSMs) used for climate research and prediction."

Consortium for International Earth Science Information Network - CIESIN
http://www.ciesin.org/

This site is the "lead-in page" and presents a large array of hypertext opportunities. "CIESIN was established in 1989 as a nonprofit, non-governmental organization to provide information that would help scientists, decision makers, and the public better understand their changing world. CIESIN specializes in global and regional network development, science data management, decision support, and training, education and technical consultation services. CIESIN is the World Data Center A(WDC-A) for Human Interactions in the Environment. The CIESIN site leads to a sister agency GCDIS.

University Corporation for Atmospheric Research - UCAR
http://www.ucar.edu/

This is a prime site for finding information, for there are links to books, articles, newsletters, lectures etc. Another "must" site for information seekers.

The U.S. Global Change Data and Information System - GCDIS
http://gcrio.ciesin.org/GCDIS/iplan/tocgcdip.html

This site is for *The U.S. Global Change Data and Information System Implementation Plan; A Report by the Committee on Environmental and Natural Resources Research, 1994*; a clickable text of interest.

Global Change: Electronic Edition
http://www.globalchange.org/

This is a "Review of Climate Change and Ozone Depletion" and it presents references to all manner of news or other information presentations; for example there are 13 pages of listings of items from September 1996: < http://www.globalchange.org/infoall/96sep1d.html > *Global Change* is published by the Pacific Institute for Studies in Development, Environment, and Security.

IEA Greenhouse Gas R&D Programme
http://www.ieagreen.org.uk/

This is a major program, based in the UK and sponsored by the IEA (operates within the OECD framework). Since 1991 the program has been evaluating technical solutions to combat greenhouse gas emissions from the combustion of fossil fuels for power generation. This organization will become the sanctioning organization for the International Conference on Carbon Dioxide Removal (ICCDR). There have been three such conferences, ICCER-3 having been held at MIT, from September 9-11, 1996, in Cambridge, MA. In the future these conferences will be called the International Conference on Greenhouse Gas Control Technologies (GHGT). The next conference will be organized by ABB Asea Brown Boveri and will be held August 31-September 2, 1998, in Interlaken, Switzerland. This site provides a portal to detailed technical discussion.

THE URGENT QUESTION: HOW CAN THE MACRO-ENGINEERING COMMUNITY RESPOND TO GLOBAL CLIMATE CHANGE?

This is a question to which members, and members-to-be, of the macro-engineering community will provide an answer. One hopes that there will not be confusion regarding the word "macro", for it can mean "being large" (such as Henry Ford's River Rouge plant); "being exceptionally prominent" (such as IBM in its mainframe days); involving or intended for use in "large quantities" (Microsoft), or on a "large scale" (communication or transportation networks). As Karl Popper decided as a young lad, arguments over the meanings of words were of little use. He decided that the only useful question was: What is the problem?

With global warming and climate change a possibility, we may be challenged by global problems, as many believe, and "macro", as in "macro-engineering" may then be challenged to take on a new meaning. But for the macro-engineering community the challenge will be the new problems that will confront them. There is no reason to believe that the macro-engineering community will not meet this challenge, especially as it will receive the informed support of an informed public and better informed governments. The fact that this is the first time in known history that all the world's people are being challenged will ensure a new historical role for the engineer.

NOTES

1. Gaskell, T.F. *The Gulf Stream*. New York: New American Library, 1972, p. 151.

2. Ibid., p. 151f.

3. Ibid., p. 152.

4. Ibid., p. 156.

5. Hughes, R. *The Shock of the New: Art and the Century of Change*, updated and enlarged edition. London: Thames & Hudson, 1991.

6. *World's Fair, Chicago 1893, The Columbia Gallery*. Chicago,: The Werner Company, 1894.

7. Gaskell, op. cit., p. 166.

8. Ibid., p. 152.

9. Borisov, P.M. *Can Man Change the Climate?* Moscow: Progress Publishers, 1973. Translated by F. Levinson.

10. Borisov's first paper on this subject was called "Radical Improvement of Climate in the Planers Polar and Moderate Latitudes". Borisov tells us that this paper was filed at the USSR Committee for Inventions and Discoveries of the USSR Council of Ministers under No. 7337 in 1957. It would be interesting to know how this suggestion was received.

11. Nevertheless, the back of the cover contains this text: "This book examines the ways to solve the very important problem of improving the climate and reducing agriculture's dependence on Nature to a minimum.... The reader will learn about the suggested projects for reforming Nature and the possibilities of putting them into effect."

12. Borisov, in his chapter "First Stages of Climatic Amelioration" (pages 137-149) writes: "The climate outside the USSR -- in Western Europe, Mongolia, Northern China and Japan -- would also improve. The recession of the Canadian anticyclone, which prevails in North America, would ameliorate the climate over the greater part of that continent. The permafrost, which extends to the southernmost tips of Hudson Bay, would disappear in the upper level...."

13. Gaskell, op. cit., p. 152.

14. Ibid., p. 158.

15. Higuchi, K., "A Possibility of Constructing a Dam to Change the General Oceanic Circulation," Paper presented at the Second International Future Research Conference, Kyoto, Japan, 1970.

16. Takano, K., "General circulation in the global ocean", *Journ. Oceanogr. Soc. Japan*, Vol. 25, 1969, pp. 48-50. In Japanese.

17. Manabe, S. & Bryan, K. "Climate and the Ocean Circulation," *Monthly Weather Review*, Vol. 97, 1969, pp. 739-827.

18. Arakawa, A. Private communication, 1968.

19. Ley, W., *Engineers' Dreams*. New York: The Viking Press, 1954.
Ley's book described "dreams" that are being realized today: the production of power from the wind and from solar radiation.

20. See: Gaskell, op. cit., p. 159 and Ley, op. cit., pp. 138-156.

21. Shortly after writing the above I received a copy of an article by Richard Brook Cathcart, "Mitigative Anthropogeomorphology: A revised 'plan' for the Mediterranean Sea Basin and the Sahara," *Terra Nova; The European Journal of Geosciences*, Nov./Dec., 1995, pp. 636-40. The author writes: "The present author has slightly modernized Sörgel's gift to *Homo sapiens*. Construction of 'Atlantropa' would provide macro-engineering and terraforming with a much-needed practice run for the remaking of our solar system's planets."

22. See Ley, op. cit., pp. 122-136.

23. Ibid., pp. 98-118.

24. Gaskell, op. cit., pp. 165-66.

25. Dubos, R. *The Dreams of Reason, Science and Utopia*. New York: Columbia University Press, 1961.

26. Somerville, R.C.J. *The Forgiving Air: Understanding Environmental Change*. University of California Press, 1996, p. 153.

27. Modern astronomy has downgraded our sun, for it is not the center of our galaxy but lies in one of its spiral arms, 25,000 light years from its center. Our galaxy is about 100,000 light years in diameter.

28. The author thanks Kathy Crosby, of The Vermont Book Shop in Middlebury, for supplying data regarding book titles.

29. There is a revolutionary view of the nature and role of earthbound life: "Over the ages, the actions of a hidden biosphere of microbes may have formed the planet's skin" and Professor Carl Woese claims that microorganisms are in charge of almost everything, stating: "If microbial life were to disappear, that would be it -- instant death for the planet." "Science Times", *The New York Times*, October 15, 1996.

30. Just after writing this I saw these tabloid headlines: *Weekly World News*: U.S. WEATHER FORECASTERS WARN: WORST WINTER IN 100 YEARS! Record cold will turn America into an arctic wasteland!" *The Sun*: "BIBLE WARNS: THOUSANDS TO DIE IN WORST-EVER FALL.

31. Jardine, L. *Worldly Goods: A New History of the Renaissance*. London, Macmillan, 1996.

32. Thompson, D. *The End of Time; Faith and Fear in the Shadow of the Millennium.* London: Sinclair-Stevenson, 1996.
The positive review of this book, in London's *Financial Times* of September 14-15, 1996 (p. XV., is titled: "Beware Cosmic Cranks".

33. Forrester, J.W. "Managerial Decision Making", In: *Computers and the World of the Future,* MIT Press, 1962, p 37.

34. Simon, H.A. *The Shape of Automation for Men and Management.* New York: Harper and Row, 1963, p. 53f.

35. Ibid., p. 88.

36. Scott Morton, M.S. *Management Decision Systems; Computer-Based Support for Decision Making,* Division of Research, Graduate School of Business Administration, Harvard University, Boston, 1971.

37. Simon, H.a. *The Sciences of the Artificial.* Cambridge, MA: The MIT Press, second printing, 1982.

38. Newall, A. & Simon, H.A. *Human Problem Solving.* New York: Prentice Hall, 1972.
A problem is defined on page 72: "A person is confronted by a problem when he wants something and does not know immediately what series of actions he can perform to get it." There may not be any.

39. In his memoirs, the Nobel chemist, James Conant, remarks that it was a blessing that engineers designed the first U.S. wartime nuclear reactors -- rather than scientists, for the engineers insisted on a conservatism in design that the scientists thought unnecessary. But Conant claimed that this approach was essential to allowing future design problems to be corrected. The role of the scientists, such as Fermi, was essential and so, Conant claimed, was the role of the DuPont engineers.

40. Marland, G., "Could We/Should We Engineer the Earth's Climate?", *Climatic Change*, Vol. 33, No. 3, July 1996.

41. "Geoengineering", Marland points out, was introduced by C. Marchetti in 1976. Marland defines the term thus: "We use the term 'geoengineering' to describe the idea of large-scale, intentional engineering of our environment for the primary purpose of controlling or counteracting changes in the chemistry of the atmosphere."

42. Table of contents of the journal *Climatic Change*:
gopher://Gopher.wkap.nl:70/00gopher_root1 %3a %5bjournal.clim %5dclim .toc

43. "Policy Implications of Greenhouse Warming: Mitigation, Adaptation, and the Science Base", a report produced by the Panel on Policy Implications of Greenhouse Warming, National Academy of Sciences, National Academy of Engineering, Institute of Medicine, ISBN 0-309-04386-7, National Academy Press, Washington, 1992.

44. "Policy Implications of Greenhouse Warming", produced by the Panel named in Note 43, National Academy Press, Washington, 1991.

6

Building Faster to Conserve More:
A Sustainable Global City

William H. Small
The Bechtel Corporation

The ability to supply the world with adequate infrastructure is one of the key issues facing the engineering/construction industry today. There is considerable concern that community and infrastructure development is not keeping pace with the needs of the world's expanding population. In fact, news reports are often marked by stories decrying the quality of life worldwide.

Is it not the responsibility of the engineering/construction industry, which can include both the public and private development sectors, to provide this critical infrastructure? And if so, why has the industry fallen behind in such important work? There do seem to be some hopeful signs that infrastructure can be developed more quickly and with less negative impact on the environment.

The lack of critical infrastructure development can be attributed in part to the prevailing belief that development automatically translates into destruction of our natural environment. The daily news is filled with warnings about the dire consequences of development, such as loss of rain forests; wildlife habitat and the ozone layer; the rapid consumption of non-renewable resources; and the water, air, and land pollution that results from excessive consumption. In our desire to meet the world's economic needs, are we approaching the limits of what our planet's natural systems can absorb?

The debates about whether to develop infrastructure or conserve ecological resources often result in long delays and large cost increases in implementing even the most well-intentioned projects. The development of new airports illustrates this point well. A modern airport is essential to the economic well-being of every large city. State-of-the-art air traffic control equipment promotes safety. High-capacity airfields and access roads reduce in-the-air and on-the-ground congestion with dramatic reductions in time, operating costs, and fuel burned.

Yet for all of these and other benefits, the new Munich Airport which opened in 1992, the new Osaka Airport which opened in 1994, the new Denver Airport which opened in 1995, and the new Hong Kong Airport which will open in 1998, each took, or will take, two decades to go from inception to commercial operations. Why does it take a generation to build a new airport?

As population increases and progress rolls on, natural resources are diminished. We seem locked in endless debate.

Governments have been the traditional developers of large infrastructure projects; they had the deep pockets necessary to tolerate the uncertainties associated with protracted debate. Today, however, governments in both the developed and developing countries are spending their limited funds on social programs, and often lack the money and political support to engage in major infrastructure development programs.

Accordingly, governments are looking more to the private sector to develop critical infrastructure, including roads, transit systems, airports, electric and water systems, and even housing. But how can the private sector successfully participate in the face of endless public debate, delayed schedules, and rising costs? And how can our industry overcome the perception on the part of the general public that private developers are driven by greed rather than community service?

The lack of adequate community facilities is a source of social unrest in many parts of the developed and developing world, including the United States. Poverty, natural resource destruction, and high population growth form a vicious cycle. If the problem continues to worsen, is there not real concern about our social fabric? Is this really the way we want to approach the future?

Our industry, both private and public sectors, has the skills and resources to provide the world with a higher quality of life. It is our business, perhaps even our obligation, to do so. We are here to meet both public and private interests, but it takes us years, sometimes decades, to build necessary, large-scale infrastructure. What causes this delay?

DEFINING THE DILEMMA
Could it be a lack of design and construction know-how? It is understandable why the super-projects of the past took so long, but today we have the design and construction know-how to complete most complex civil projects in five years.

Is it a lack of financial resources? Sometimes this is a problem, but today development money is generally available for financially viable projects. Government funds, loans from multilateral lending institutions, and private local and overseas equity can be blended into packages for all sorts of infrastructure projects, from water systems to telecommunications networks.

It appears that the cause of the long cycle time is institutional in nature. The delay is caused because we have not been able to resolve our need to conserve natural resources with our need to develop them. The conflict in value systems has created a debate that has been difficult to resolve because both needs -- and the value systems that drive them -- are equally valid.

Organizational psychologists use the Johari chart to describe the relationship between two competing values sets. Exhibit 6.1 plots development versus conservation. On both axes, value is measured as the degree to which either development or conservation wins the debate regarding project go-ahead.

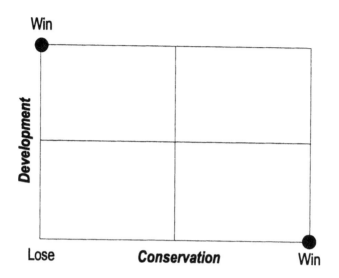

Ex. 6.1 Development vs. conservation

A dot in the upper left-hand corner indicates that development always wins the debate over conservation concerns. A dot in the lower right-hand corner indicates the opposite -- that conservation concerns always thwart development. These are extreme positions and represent equally unsatisfactory conclusions. Too much development without a proper concern for the environment produces congested and polluted cities. On the other hand, no development at all results in a declining economy and associated quality of life - an equally unsatisfactory condition.

Both value systems have merit. The proponents of development are not conspiring to destroy the ozone layer or rain forests. Likewise, conservationists are not against good jobs and transportation. The debate concerns everyone in search of satisfactory solutions.

A HISTORICAL PERSPECTIVE
How did this conflict between development and conservation arise?

As we approach the beginning of the third millennium, it is interesting to look back on the last two. As illustrated in Exhibit 6.2, man has lived in balance with nature for the vast majority of the last 2,000 years.

During most of this time, man lived as hunter/gatherers or as simple farmers -- many people of the world still do. Both economic orders are land-intensive and require that people live close to their food supply. Generally, trees are the fuel of choice. Wood and stone are the building materials of choice. These lifestyles inherently promote heavy consumption of environmental resources. But because population and technology were limited during most of the first two millennia on the timeline, the impact on the world's resources was small and people lived at an environmentally sustainable level.

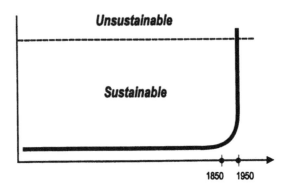

Ex. 6.2 Living in balance

The Industrial Revolution gained a full head of steam in the mid-1800s. It permitted the intense consumption of coal for power and the making of steel. This gave birth to the large-scale exploitation of mineral resources and clearing of land. As a result, large industrial cities emerged with crowded tenements, smoke stacks belching pollutants, and open sewers flowing into rivers.

The rise of the services economy in the middle of the 20th century created greater mobility for people, and urban development spread beyond the cities. Freeways, airports, and suburban housing and industrial estates took more land.

By the 1960s -- just a hundred years after the beginning of the Industrial Revolution -- the consumption of natural resources to support our growing world population and expanding lifestyles emerged as an issue. We began to gain a scientific understanding of the limits to which the earth could absorb human activities. We saw the first public debates over the consequences of development and the introduction of environmental regulations to control development.

Almost universally, the issue was viewed as a zero sum game in which one side won at the expense of the other. The pursuit of progress was at the expense of the natural environment. A call to preserve the environment was almost always also a call to halt further development. Activist groups took shape and language on both sides became emotionally charged.

Exhibit 6.3 shows this confrontation. A dot in the middle of the Johari chart represents the alternating "I win, you lose" and "You win, I lose" scenarios. This is often what happens today when issues must be resolved in the courts. Inevitably, extended delays and increased costs result, and the cycle of poverty and environmental degradation continues.

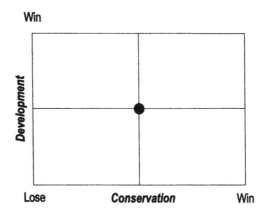

Win

Lose   *Conservation*   Win

Ex. 6.3 A "win-lose" scenario

Continued technological advancements, however, are changing the basic nature of our economy. Trends are emerging which suggest that we may be evolving beyond the stalemate between economic development and conservation of natural resources.

THE KNOWLEDGE WORKER

In today's digital age, a new dominant worker is emerging -- the knowledge worker. The rise of the knowledge worker represents a fundamental shift in human values and with it a fundamental change in how we interact with our environment. The knowledge worker can be defined as someone whose job simultaneously involves producing and learning. Their work generates new knowledge which is used to enhance the value of their work. Given the rapid pace of technology, most knowledge workers must renew their professional knowledge every two to four years.

There is a significant contrast between knowledge workers and industrial and service workers who perform the same task over and over. These workers may be very skilled and become more skilled over time, but they are not learning on the job in such a way that they fundamentally change their jobs on a continuing basis.

What are the characteristics that make the knowledge worker different from past workers? Knowledge workers' lives are inherently different. They intermingle the activities of home, work, school, health care, recreation, and culture. This lifestyle favors compact land use -- an urban setting. Suburban sprawl is not supportive of these values.

Many of today's large universities provide this kind of integrated environment. Some of today's best cities offer residents the ability to move quickly and easily between home, work, school, shopping, and recreation.

As knowledge workers are both male and female and pursue active careers, they tend to have fewer children and to have them later in life. Knowledge workers are better able to provide economically for their families; a large family provides no economic benefit in the digital economy.

Knowledge workers value a high quality of life, free from congestion, noise, and air and water pollution. They will accept the personal choices necessary -- such as the use of green consumer goods, public transportation, energy conservation and recycling -- for achieving a sustainable environment.

Knowledge workers use computers, pagers, cellular phones, and fax machines for personal and business tasks. They shop, bank, visit with friends, even take mini-vacations via electronic networks, reducing the need for routine travel, thus saving fuel and reducing air and noise pollution.

Because knowledge workers are well-educated, they understand the need to care about and plan for the future. They actively and willingly support the concepts of compact cities integrated with open space, of smaller families that lead to stable populations, of energy and wildlife habitat conservation, public transportation, recycling, and clean water, air, and land. They understand the need to live today in such a way that their children's ability to enjoy tomorrow is not compromised.

The knowledge worker lifestyle is the scenario that characterizes the sustainable global city. Many cities already contain aspects of sustainability. Singapore and Hong Kong in Asia Pacific, London and Zurich in Europe, and San Francisco and New York here in North America are examples that show promise. In the future, some of the world's fully sustainable cities will be existing ones rebuilt to reflect the new values; others will be built from the ground up.

The rise of knowledge workers and their sustainable global cities also offers us the opportunity to close the infrastructure gap and provide people with the healthy living conditions that everyone deserves.

The rise of knowledge workers also makes it economically, socially, and politically possible to create sustainable global cities. Hence the faster we can create knowledge workers and their sustainable global cities, the faster we can realize the conservation benefits that their value system supports -- and a return to the sustainable zone illustrated in Exhibit 6.2.

SEIZING THE OPPORTUNITY

The rise of knowledge workers and their sustainable global cities provides the opportunity to revise the current equation from one of confrontation between development and conservation to one of partnership. By finding mutually satisfying solutions, both sides can win. This is illustrated in Exhibit 6.4; the dot in the upper-right corner of the chart indicates a "win-win" scenario.

But how do we get traditionally adversarial sides to work together? How can we assure conservationists that infrastructure projects will actually advance the arrival of knowledge workers and ultimately help to conserve the environment? Similarly, how can we assure private developers that their investments will not become bogged downed in endless debate?

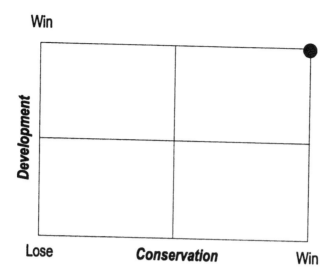

Ex. 6.4 A "win-win" scenario

The knowledge worker is again the key to the solution. Their simulation technologies are progressing to the point where visualizing the future is commonplace. We are reaching the point where we can understand the consequences of a specific development before the site is physically touched.

Simulation is giving us the ability to calculate impacts accurately, effectively communicate information to all interested parties, quickly incorporate design changes, and monitor compliance with plans during implementation. Simulation technologies offer the opportunity greatly to increase confidence in the project approval process and, ideally, lead to less debate and fewer schedule delays and cost increases.

Projecting into the future, technology will allow us to predict with even greater accuracy the effects of development on natural environments all over the world, and their resultant effects on population growth. Simulation technology is developing so rapidly that it may emerge as a new field servicing both public policy and private development.

The potential for new partnerships based on an ability to work out critical differences between disparate interests through simulation is significant. This approach does not suggest that either value system be compromised. Rather, it offers mutually satisfying outcomes.

Combining these concepts of accelerating the rise of the knowledge worker and using simulation technologies to build consensus about how development should be implemented raises the interesting notion that building faster will achieve our conservation goals sooner than is currently possible.

Imagine the progress that could be achieved if the opposing forces of development and conservation could be realigned to work together. Instead of offsetting each other or canceling each other out, the energies would be additive. How much momentum could be achieved if all parties pursued a common goal?

CONCLUSION
The creation of sustainable global cities is essential if we are to conserve our limited natural resources and provide people with the healthy environment that they deserve. The rise of knowledge workers, and their attendant value system, creates the economic, social, and political force that makes this picture of the future possible.

The engineering/construction industry can be at the forefront of exploring ways in which simulation technologies can build powerful new partnerships to resolve the dilemma of development versus conservation. The pursuit of these ideals can ultimately lead to the emergence of truly sustainable global cities.

7

A New Urban Model

Donna Goodman
Architect, AIA, New York City

Historians claim that Nero burned Rome, so he could later rebuild the city in his own image. It was a drastic approach to urban renewal, even for Nero. But he was not alone in his desire to transform the urban environment. Throughout history, architects and engineers have designed alternative concepts for cities.

In the late 19th century, Tony Garnier created an industrial city with numerous community facilities. Several decades later, Frank Lloyd Wright proposed an agrarian society, rejecting urban life. In the early 20th century, Le Courbusier developed a city of skyscrapers and open fields, trying to integrate nature with the urban environment.

Each architect presented a physical plan of a city, but in each design, a larger philosophy of social planning was expressed. Gamier's *City Industrielle* expressed Fourier's socialist-utopian vision. Wright's Greenacre City housed Emerson's notion of individualism. Le Courbusier's *Ville Radieuse* echoed Bauhaus theories logical and efficient planning. The struggle among early modernists to design cities that supported social values created an atmosphere of idealism and experiment. Each of the projects had faults, but the dialogue generated was valuable. In recent years, very few projects of this kind have been developed, though the need for new urban theory has never been more profound.

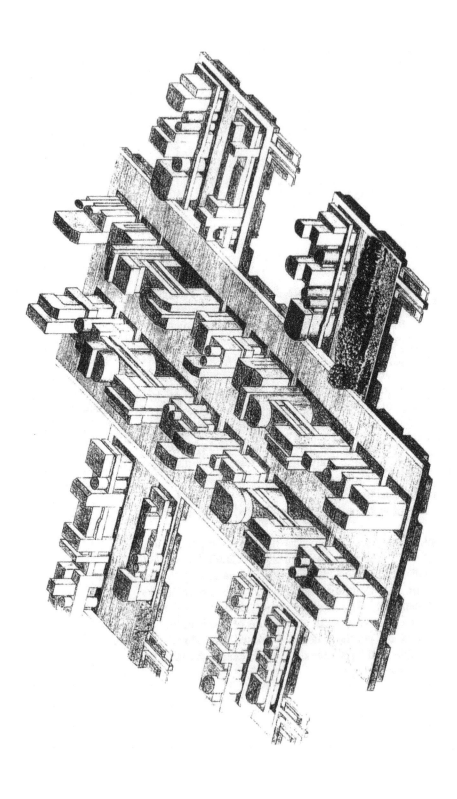

THE CURRENT URBAN DILEMMA
The world population is expected to double in thirty years. Twice as many people will need food, housing, energy, and water. Demand for jobs, education, and medicine will also double. According to a recent study by The World Bank, most growth of the will occur in third world cities which have few resources. Some African cities are already growing 10% a year.

Most developing cities must begin with infrastructure. In China, two hundred million people still need electricity. In Mexico City, one-third of the population has no water or sewage systems. These cities will also need extensive educational systems to introduce technology and skills.

Developed cities are also struggling with growth. Some cities in Japan are approaching twenty million people. European cities are fighting urban sprawl. Cities like New York are facing increasing costs. Although New York's population is almost stable, the costs of city services have increased and public systems are being reduced.

SURVEY OF SOLUTIONS
In preparation for growth, architects are discussing the development of urban regional corridors and disbursed urban minicenters. Europeans are building new rapid transit systems. Americans are considering new subways and infrastructure, and Japanese cities are expanding offshore. Over sixty artificial islands have been built for housing, transportation and urban expansion.

Cities in developing countries are creating more modest proposals. An innovative mayor in Curitiba, Brazil has designed a new housing concept. He integrates low and middle-income families in the same neighborhoods to avoid creating slums. In Cairo, an informal system of recycling employs many poor families.

New farming and industrial technology will also be important. Nearly 40% of our resources will come from the ocean by the mid-21st century. Europeans are creating ocean energy, Americans are exploring ocean mines, and the Chinese have cultivated over two million acres of aquaculture.

The transition from fishing to aquaculture will be as profoundly important to future generations as the shift from hunting and gathering to farming was to social development in the past. Hundreds of products have already been developed from aquaculture, including food, drugs, pharmaceuticals, paper, and fuel.

The next generation of designers and planners will face new issues. The scale of cities has grown. The issues have expanded to include technology, environment, and education. The objective has changed from designing an "ideal city," to creating a flexible urban environment, able to evolve in an age of rapid change.

A NEW URBAN MODEL: PLANNING AND DESIGN CONCEPTS
The working title of this model is *The New Urban Anatomy*. The concept includes new design and planning concepts; social, political and economic proposals; and environmental systems and offshore designs. The ideas are presented through the design of a conceptual future city which integrates these concepts into a single proposal. The project includes:

A Flexible Master Plan: At the height of the empire, Rome had 1.1 million people. Some modern cities will approach thirty million people by the end of the decade. In this conceptual city, areas for flexibility and growth are built into the master plan. Streets on the waterfront contain flexible structures that can be moved or changed to expand an industry, transform a neighborhood, or eliminate an obsolete function. Flexible waterfront zoning also supports seasonal events or temporary visitors to the city, such as a university ship, a tourist barge, or an offshore park.

Vertical Zoning: Most cities have a horizontal master plan. This future city also has an intricate vertical organization to support new functions and environmental goals. The zoning includes new underground services, street level functions, and roof and upper-level recreational activities.

Streets for Professions: The mere mention of Broadway conjures up visions of theaters and neon lights. The image of Wall Street is financial power. Lewis Mumford called this phenomenon "spontaneous grouping of professions," a practice as old as Ninevah and Ur. In this future city, professional communities play an important role. Certain streets are designed for specific professions, such as a "street of the arts" or a "medical street." The design of the street reflects the character of activity. Grouping of professions is physically convenient and helps strengthen dialogue in the city.

Fig. A new urban model

Flexible Buildings: There was a time when a house remained a house and a factory remained a factory. Permanent arrangements of space are no longer possible. Here, the design of open spaces, combined with flexible building systems, allows occupants to alter their interiors without major construction. An inventory of kitchens, bathrooms, closets, and wall units are available. Rent is determined by interior plans.

Regional Planning: In this city, urban growth is anticipated by the development of regional corridors of transportation and services. The corridors help to distribute population along planned areas of growth. In each "linear city," a new minicenter is built to provide an urban focus area. Offshore concepts also provide a direction for initial expansion of the city. They include piers, flexible areas, and artificial islands.

SOCIAL, POLITICAL AND ECONOMIC SYSTEMS
The city's philosophical goal is to support individual development. This is accomplished by creating smaller communities within the city, providing new housing, education, and employment options, and integrating industry and university systems to support growth.

Town Square as Information Center: When Marco Polo went to China, he had little support in his efforts to become familiar with their customs. In this city, tourist information booths offer multilingual information on the master plan and urban resources of the city. For long-term residents, the main streets of the city have information centers which offer global access to jobs, education, and markets. The information centers also provide an introduction to the urban master plan exploring neighborhoods, transportation, housing, parks, businesses, and schools.

Teleconference: In the 1992 U.S. presidential election, Ross Perot made "video town meetings" a common idea. Never before had a candidate used media more actively to engage the public. In this proposed city, teleconference centers are available for business meetings, educational conferences, or personal conversations and events. Facilities of various sizes can be rented. The city also uses video meetings to provide linkage to other urban centers and expand the dialogue on political issues.

Multiple Economic Systems: More economic experiments occurred in the 20th century than in any other era in history. New forms of communism, capitalism, and socialism were developed. In this city, individuals choose from several economic systems. Capitalism is the primary system, but there are also cooperative communities for students, elderly, or struggling families. The cooperatives are linked to industry and offer part time work, housing, daycare, job training, and support systems.

Political Technocracy: The first Congress had doctors, farmers, industrialists, and inventors. We still need the expertise of the entire society. In this city, the professions play a large role in government. One House of Congress is recreated as a cross-section of representatives from trades and professions. Members are elected by individuals in field. Groups like housewives, elderly, or minorities are also represented. The new system reduces lobbying and brings more variety of talent to government.

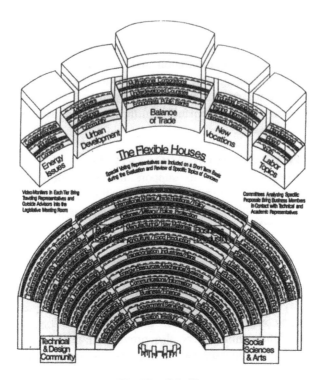

The Flexible Houses

ENVIRONMENTAL SYSTEMS AND OFFSHORE DESIGNS

Environmental issues are addressed throughout the project. The city uses clean resources. Landscape, planting, and orientation reduces the energy demand and improves air flow. Public transit, recycling, and environmental systems are built throughout the city.

The Urban Underground: New York City mail was once moved through underground tubes. In Moscow, some types of trash were carried in vacuum pipelines. In this city, all deliveries are placed in new underground systems, removing trucks and services from main streets. Subways are built on the upper tier of the underground. On a lower tier, underground conveyors carry mail, deliveries, transportation, and trash, and are linked to basements of buildings. Infrastructure is also built into the tunnels, providing easy access for repairs.

Roof Parks: At roof level, recreational and cultural activities are integrated into block-long roof parks, planted terraces, and upper level streets. These roof parks provide additional green space, desirable views, and other amenities.

Some roof areas contain infrastructure, such as satellite systems, to monitor environmental conditions and enforce new environmental laws and standards.

The Waterfront: The waterfront is a major economic resource for the city. Many activities are located on the waterfront to alleviate traffic in the urban center.   Piers are created for major commercial and recreational activities. Neighborhoods with flexible zoning supports industry, or visiting business and tourist groups.   Water transportation reduces stress on other forms of urban transportation.

8

World Water Supply Issues

Ernst G. Frankel
Professor of Ocean Engineering, M.I.T.

Fresh water has long been assumed to be renewable and a resource readily available in most parts of the earth. It is only in recent years that we have come to recognize that it is also finite. As world population doubles again in the next 40 years to nearly 10 billion, and per capita water consumption levels increase, water scarcity will intensify. While earth seems to be largely occupied by water (73.1% of the surface by ocean and 1.3% by fresh water lakes), only about 2% to 5% of all the water on earth is fresh water, and over 66% of this is locked in glaciers and ice caps. Only about 110,000 cubic Km/year or 0.000008% of all water on earth is renewed every year by the solar hydrologic cycle. This scenario will not change much, and therefore the combined effects of population growth and per capita consumption pose a real threat.

About one-third of the renewable fresh water supply (40,000 cubic Km/year) is fresh water runoff from land to sea. Since 75% of runoff is flood water, dams have been built that collectively hold 14% of all the runoff. This contributes to a stable supply of water provided by underground aquifers and year-round river flows. As a result, the total stable renewable supply is about 14,600 cubic Km/year, of which an estimated 12,500 cubic Km/year are currently accessible for irrigation, industrial, and human consumption.

Global fresh water consumption in 1990 was 4,430 cubic Km/year or 35% of accessible supply. Another 20% was used for in-stream activities such as pollution dilution, transportation, etc. In other words, we now use over 50% of the accessible fresh water supply, more than three times the consumption in 1950. Obviously we cannot again triple our water consumption in the next 40 years. Something therefore has to give.

Global runoff as a percentage of population is quite unevenly distributed, with the ratio of percentage of runoff to percentage of population varying as follows:

4.166 for South America
1.875 for North and Central America
0.846 for Africa
0.615 for Europe, and
0.600 for Asia.

While South America has an abundance of water, Asia and Europe, which together account for 73% of the world's population, are water-poor. Until 1970 over 1,000 large new dams were brought into operation every year. This number has now dropped to below 250 per year. It is becoming increasingly expensive and difficult to increase accessible runoff.

For years desalination of ocean water was hailed as the potential saving technology, but even today the cost of desalination is still too high both in terms of monetary and environmental impact. Current costs are $1.00-1.60 per cubic meter. With over 11,000 desalination plants installed (7.4 billion cubic meters/year), this still accounts for less than 1% of world water use.

Water constraints are usually estimated by comparing renewable water supply to population size. If renewable run off drops below 1,700 cubic meters per person per year, a country is considered short of water, and if this number falls below 1,000 cubic meters per person per year, it is considered water scarce. The increasing water scarcity in the Middle East, Africa, and Europe is bound to have major consequences on food production, ecological and social support systems, and political as well as strategic positions. This problem appears to be more effectively managed in Asia, notwithstanding its low ratio of accessible water to population density. Countries such as Algeria, Egypt, Libya, Morocco, Tunisia, Israel, Jordan, and the whole of the Arab Peninsula also fall into this category. In 1995 44 countries with a population of 733 million were short of water and the number is growing. Exhibit 8.1 shows African and Middle Eastern countries currently dealing with water shortages.

Another important issue is ground water depletion which is particularly severe in North Africa, Israel, and North China. The southwestern U.S., Arabian Peninsula, and northwest India (Punjab, Gujarat, etc.) are other areas with severe groundwater depletion. All of this is affecting irrigation, and the irrigated land area (per 1000 people) has fallen steadily from 48 ha/1000 people in 1975 to 44 ha/1000 people in 1995.

Another increasingly difficult issue is the reallocation of water from agricultural use to cities. In China in particular water supplies are increasingly being taken from farm lands to cities by overpumping ground water.

The growth of mega cities with populations in excess of 10 million is a particular problem. While in 1975 only 12 cities had populations larger than 10 million while worldwide the number has since grown to 25 and is expected to grow to 50 by 1020.

Consider next the effect of large dams whose number has grown from 5,000 to over 38,000 since 1950. The most important concern is the effect on silt and nutrient flow into the oceans and a resulting regression of river deltas. There are also

Country	Internal Runoff Per Capita 1995	1995 Population	Projected 2025 Population
AFRICA			
Algeria	489	28.4	47.2
Burkina Faso	1,683	10.4	20.9
Burundi	563	6.4	13.5
Cape Verde	750	0.4	0.7
Djibouti	500	0.6	1.1
Egypt	29	61.9	97.9
Eritrea	800	3.5	7.0
Kenya	714	28.3	63.6
Libya	115	5.2	14.4
Mauritania	174	2.3	4.4
Morocco	1,027	29.2	47.4
Niger	380	9.2	22.4
Rwanda	808	7.8	12.8
Somalia	645	9.3	21.3
South Africa	1,030	43.5	70.1
Sudan	1,246	28.1	58.4
Tunisia	393	8.9	13.3
Zimbabwe	1,248	11.3	19.6
MIDDLE EAST	1,650	20.6	52.6
Iraq	309	5.5	8.0
Israel	249	4.1	8.3
Jordan	0	1.5	3.6
Kuwait	1,297	3.7	6.1
Lebanon	909	2.2	6.0
Oman	119	18.5	48.2
Saudi Arabia	517	14.7	33.5
Syria	158	1.9	3.0
UAE	189	13.2	34.5
Yemen			
Total 1995		380.6	739.8
Additional Countries by 2025			618.0
Projected Total 2025			1,357.8

Source: "Dividing the Waters," World Watch Paper 132, Sept. 1996.

increasing political and strategic issues involved as more and more countries depend on imported surface water, usually rivers that originate elsewhere. Countries such as Egypt, Hungary, Mauritania, Botswana, Bulgaria, the Netherlands, Syria, Cambodia, Sudan, and others all depend on 80% or more of their water supplies on imported surface water.

As these issues continue to mount, different water management options will have to be implemented, as water management schemes are always more efficient than development of new water sources. As noted, 4,430 cubic Km/year is the current world demand, of which agriculture accounts for 2,880 cubic Km/year, industry for 975 cubic Km/year, and human consumption for 300 cubic Km/year. Reservoir losses are estimated at 275 cubic Km/year; this is one area where significant savings could be made. But there is also waste by consumers. While agriculture uses 82% of the water demanded, industry actually consumes only about 10% of its water demand.

Well over two hundred rivers flow through two or more countries, but there is no enforceable law on the allocation and use of such international waters. An interim agreement has been reached between Israel and the Palestinians on groundwater rights on the West Bank of the Jordan, but there are other international tensions over water in the Middle East, notably over the effects on Syria and Iraq of the Ataturk Dam being built in Turkey. In India, there is a dispute over the Farakka barrage, 18 miles upstream from Bangladesh, which claims that as a result it is now facing an environmental disaster due to desertification. Southeast Asia, North and Central Africa face similar problems.

The world community seems slow to recognize that the world water shortage and regional imbalances between water and population will become progressively more acute. It is estimated that twenty-five years into the next century some three billion people could be living in countries with acute water shortage. To avoid this explosive situation with all its effects on food production and natural systems in general, and the resulting threats to regional stability and peace, is a pressing macro-problem to be addressed with urgency.

9

A Rhone-Algeria Aqueduct

Ernst G. Frankel
Professor of Ocean Engineering, M.I.T.

The Rhone River is the only major river/watershed in France draining into the Mediterranean Sea. It constitutes the largest fresh water inflow into the Western Mediterranean, and suggestions have been made that it could provide a substantial fresh water supply to arid areas of Algeria south of the Atlas Mountains which border the Sahara Desert.

We have undertaken studies to determine the feasibility and costs of pumping some of the Rhone River effluent to Algeria and across the Atlas Mountains and then compared these costs (investment and operating or delivery costs) with those of generating similar fresh water supplies by distillation of sea water. Costs assumed in this study are world prices such as for construction and fuel.

Exhibit 9.1 shows the drainage basin of the Rhone River and the shortest submarine aqueduct alignment between the Rhone estuary and Algiers, including an Atlas Mountain transit and southern Algerian fresh water distribution system.

It is assumed in the study that access to the river water is at no cost to the project and that the right of way across Algeria has no cost. Similarly, no costs are included for the right of laying a submarine aqueduct in the coastal waters of France and Algiers or for the right of suspending such a pipeline in international waters. The project includes all investment and operating costs from the Rhone estuary to a central point in South Algeria without distribution costs.

The Rhone River, although short, has a large water flow because of its total fall of 5,898 feet (approx. 1,798 meters) over a length of 505 miles (812 Km).

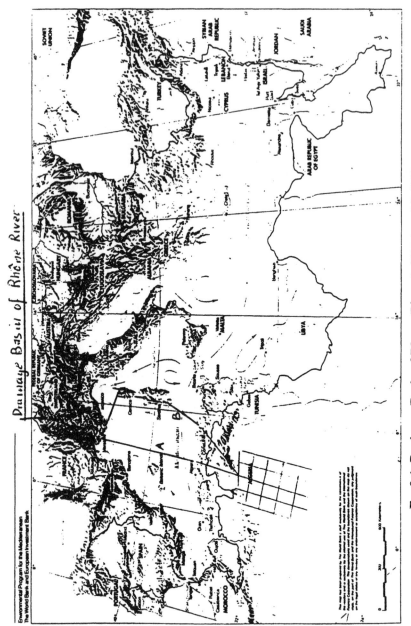

Ex. 9.1 Drainage Basin of Rhone River to Western Mediterranean

The Rhone and its tributaries form a delta starting at Arles with two principal branches. The Grand Rhone branch itself carries over 80% of the outflow, depositing its sediment in the abyssal plain of the Balearic Sea between Corsica/ Sardinia and the Spanish peninsula.

Two alternative alignments of a buoyant submerged aqueduct were investigated:
1. a shortest distance aqueduct, and
2. a shortest submerged distance aqueduct which uses Corsica and Sardinia shore corridors.

In both cases the aqueduct is designed to carry the same amount of water.

RIVER CONDITION AND WATER TAKEOFF

Half of the Rhone's potential hydroelectric power has been harnessed by an integrated series of dams, the largest of which (335,000 Kw) is at Genissia, followed by Montelima, Baix, and Andre Bloudel (near Bollene - 300,000 Kw). The river is navigable, with a minimum depth of 165 meters all the way up to Lyons. There are no dams south of Lyons. Traffic on the river carries wine, bauxite, sand, gravel, petroleum products, and fertilizers, and is the most important waterway in France in terms of volume of cargo carried.

The estuary between Arles and the shore is particularly active. To take off a major portion of the fresh water flow of the Grand Rhone before discharge into the Mediterranean Sea would require:
1. a dam with a large collector,
2. a syphon eductor, and
3. a funnel scoop.

Considering the importance of navigation on the river, a dam in the estuary would not be effective, particularly as the delta is flat. Therefore a syphon eductor or funnel scoop is proposed to be installed at a distance about 15-18 Km south of Arles where the water is still fresh, but channel depths are 3-4 meters, and thus would not to be affected by take off of 40-50% of the water flow during fall or the low water season.

The required capacity for a 40% take off in low water and 30% take off in high water season is about 1.0 billion cubic meters per year (3.00 million cubic meters or 800 million gallons per day). This requires a pipe diameter of 7-10 meters depending on valving and pumping arrangements.

It would be most desirable to configure the aqueduct system to avoid any valving or pumping at sea. This could be achieved by installing pumps at the entrance to the sea leg and have each conveyance leg of the system employ an inverted syphon effect for transporting the water.

THE AQUEDUCT

The aqueduct should be designed for a life of about 200 years and an ability to sustain all possible seismic, hydrodynamic, and other physical phenomena. The objective is to achieve optimum economy in pipeline design, construction, operation, and maintenance. Because of the underwater terrain and soil conditions on the Western Mediterranean abyssal plain, a flexible buoyant pipeline system was studied and designed, in preference to a buried pipeline.

Subsurface currents along the alignment area at depths of up to 100 meters appear to be small as are the effects of surface wave actions. Turbidity and density currents as well as turbulence along the alignment are similarly insignificant.

The underwater distance of the direct line A is 743.8 Km, while that of the indirect line B (both shown in Exhibit 9.1) is 192 Km, 72 Km, and 253 Km in three separate underwater sections, for a total underwater distance of 517 Km. To this must be added 122 Km, 132 Km, and 200 Km, or a total of 454.2 Km of overland aqueduct along the Riviera, across Corsica, and across Sardinia. Therefore, while the underwater distance is 225.8 Km shorter, more than 454 Km of additional land aqueduct would be required. On the other hand, the separate underwater aqueducts would permit several stages of booster pumps to be installed. The differences in distances in Algeria are marginal.

The pipe can be made of filament span fiberglass, precast reinforced concrete or steel. For our purposes, the thicknesses, material quantities, materials, and related cost items were identified for the three alternative materials. The aqueduct is designed to be always full and at a pressure in excess of the prevailing static head. As a result, leaks could only result in fresh water outflow but not salt water inflow.

To assure retention of the flow velocity some electric induction pumps may be considered along the submarine aqueduct. For the purposes of this conceptual feasibility study, we will assume that a reasonable flow head and velocity will be maintained using intermittent pressure or velocity inducers. These may be mechanical or hydraulic.

MATERIAL QUANTITIES AND COSTS OF AQUEDUCT PIPE

The total surface area of the direct (A) aqueduct is 23.36 million m^2 which requires the following material quantities and costs (in 1994 present value terms):

1. Steel pipe - plate   9.9 million tons
          - stiffeners   1.0 million tons
          - other   0.4 million tons
          Total   11.3 million tons

The fabricated cost of aqueduct pipe, including internal and external coating is therefore $5.65 billion. Laying costs $0.80 billion. Anchoring costs $0.40 billion. Total pipe costs $6.85 billion.

Similarly,

2. Fiberglass pipe aqueducts
      Total costs $8.28 billion

3. Reinforced concrete pipe aqueducts
      Total costs $4.98 billion

The reinforced concrete aqueduct is the least expensive and has the longest expected life, but has a greater potential for damage as a result of impact or flexure because it is more brittle and less elastic.

When alignment (B) is used, the submarine pipe lengths are significantly shorter -517.0 Km versus 743.8 Km, but a 454.2 Km set of land sections must be added. These result in the following estimated costs.

	Submarine	Land Portion	Total Cost
Steel pipe	$4.73 b	$3.69 b	$8.42 b
Fiberglass	$5.71 b	$4.46 b	$10.17 b
Reinforced concrete	$3.43 b	$2.68 b	$6.11 b

It should be noted that alignment (B) costs are approximately 21-23% more in pipe and laying costs.

INLET TAKE OFF
The inflow eductors or funnel scoops in the Rhone River with the connecting 10-20 Km buried pipe will cost approximately $200 million and should preferably be made of reinforced concrete, independent of the chosen aqueduct material, to prevent erosion and corrosion.

PUMPS, EDUCTORS AND FLOW GENERATORS
The active flow-inducing equipment requires a larger engineering effort than can be performed in this brief chapter. As a result, scaled-up estimates derived from a California aqueduct study are used. Cost of pumps, eductors, etc. including controls are $420 million.

SYSTEM COSTS
The total investment costs are estimated only to the Algerian coast, as the internal pipe and distribution systems costs would be the same if the water is extracted from the Rhone River or by distillation from the Mediterranean Sea on the Algerian coast.

Operating costs were calculated considering only the following factors (in 1994 present value terms):

1. (Fuel) pumping costs/day $ 30,000
2. Maintenance costs/day $ 40,000
3. Manning and Administrative
    costs/day   <u>$ 10,000</u>
               <u>$80,000/day</u>

Therefore total investment costs for the aqueduct will vary, as follows:

	A	B
Steel	$7.47 b	$ 9.02 b
Fiberglass	$8.90 b	$10.79 b
Reinforced Concrete	$5.60 b	$ 6.73 b

Assuming a finance cost of 6% and an amortization over a 50-year period, the financial costs per year would be:

	A	B
Steel	$373.6 m	$451.0 m
Fiberglass	$445.0 m	$539.0 m
Reinforced Concrete	$280.0 m	$336.4 m

The resulting total daily cost (including operating costs) are

	A	B
Steel	$1.10 m	$1.32 m
Fiberglass	$1.30 m	$1.56 m
Reinforced Concrete	$0.85 m	$1.00 m

In terms of cost per cubic meter of water delivered, this implies a cost of $0.28/m^3 to $0.51/m^3.

COMPARATIVE COSTS OF DISTILLING WATER DIRECTLE FROM THE SEA

The cost of distilled water depends largely on fuel costs. Using recent experience in Kuwait as an example, where a smaller 500,000 m^3/day plant is in operation, financial and operating costs (excluding fuel) are about $0.25/m^3. If fuel costs are added at world prices (and not as subsidized or waste fuel (gas)), then the cost will increase to about $0.30/m^3. Obviously, a larger plant producing 3 million m^3/day would probably have costs approximately 30-40% lower or about $0.20/m^3.

The aqueduct system is therefore marginally competitive or even attractive if:

1. the Rhone water can be extracted at no cost for the water or access,
2. there are no right of way costs for the aqueduct in coastal waters, and
3. fuel prices increase to $25/barrel and correspondingly for gas.

Then the aqueduct has a clear economic advantage over distillation.

Obviously the above study considered only the most important cost items. Differences in maintenance costs, for example, were not considered, nor were differences in pumping power and costs for alternative systems. It is believed that the above costs are within $\pm 25\%$ of investment and operating costs, all expressed in net present 1994 terms.

SUMMARY

I believe there is sufficient evidence of prospective technical and economic feasibility of a trans-Mediterranean aqueduct concept as a whole to proceed with a reconnaissance-level investigation. This opinion is subject to the following reservations:

a. The first-order calculation of critical buckling radius of the neutrally buoyant aqueduct conduit should be completed and the results compared to the magnitude of deflection anticipated in the most severe first-order hydroelastic example. Results could affect the relative economic feasibility of the buoyant conduit concept.

b. More research should be conducted on the probability of frequency and magnitude of major storms. Expected seawater velocity in the route zone during the passage of a large storm should be reassessed. The shock and hydrodynamic motion considerations of local storms should be more thoroughly reviewed.

The buckling calculations, which should be part of a decision to proceed with a reconnaissance investigation, have been done, and it is concluded that the buoyant conduit concept could absorb a substantial increase in storm-induced lateral drag forces. The second- and third-order hydroelastic analyses should include analyses that consider loads generated by hypothesized cases of seismic displacements.

Some of the more critical questions to be answered by a valid reconnaissance investigation are the following:

A. *Hydrodynamics*
1. How accurate is the hindcasting technique in ascertaining the size of gravity waves?
2. What maximum currents will be experienced?
3. What is the direction of these maximum currents?
4. What vortices may develop in these maximum currents?
5. What is the maximum velocity to be expected during the presence of internal waves?
6. What maximum storms should be considered?

B. *Bathymetry*
1. How accurate is the Geodetic Survey's bathymetric data?
2. How much change in detail contour configuration will result from correlative sonar sounding work?

C. *Geology/Soil Mechanics*
1. Will the unconsolidated sediments in the route zone be sufficiently defined by proposed intermittent fine-grained sonar survey to provide design information for the study of anchor configurations? If not, what additional survey work must be done? For what locations on the route?
2. What settling rates for practical mass weight anchors must be considered?
3. What will the static and dynamic coefficients of friction be for the optimal mass weight anchors versus an unconsolidated foundation?
4. What percentage of anchor stations should be positively secured to consolidated rock?
5. What will be the nature and extent of the drift of sediments around given anchor shapes, or against pipe laying, or partially buried in the bottom at various considered bottom depths?
6. How deep should excavation be for protectively covering a buried conduit? How much consolidated rock must be excavated?

7. Where on the route is there any appreciable energy potential in formations reposing above the route level which may generate damaging turbidity currents upon seismic excitation? What special engineering is indicated for any such locations?
8. How much displacement can be expected in a maximum earthquake along known faults? What allowances should be made for random displacement along secondary unknown faults associated with the known faults?
9. What is a valid definition of a maximum local storm?

D. *Fouling*
1. What is the rate of growth of predominant fouling life at representative points along the route zone?
2. What maximum growth density of the above may be expected during the life of the aqueduct?
3. What internal fouling may be encountered? Can it be controlled with additives and/or filtration at the headbay reservoir?

E. *Hydroelastic Analyses*
1. All applicable questions presented in second order analysis discussed in this report.
2. After collection of survey data, all questions applicable in third order analyses as described in this report.

F. *Analyses of Sum of Forces*
1. What is the maximum bending moment, deflection, tensile stress, compressive stress, and shear stress for a full range of considered pipe cross-sections and material properties when hydroelastic maxima are combined with steady state conditions as applicable?
2. Do any of these values indicate buckling either when the aqueduct is pressurized for water conveyance or with little internal pressurization during installation and maintenance?
3. What cumulative forces act on the pipe and retention members in selected cases of pipeline route configuration?

G. *Hydraulic and Maintenance Analyses*
1. Are laboratory experiments required for establishing drag coefficients? If so, what are C_{Ds} and C_{Du}?
2. Can water hammer phenomena be generated within the pipe by a seismic event? What is the value of any resultant forces in the pipe structure?
3. Will flexural waves in the pipe induced by high sea conditions result in excessive head loss or surging? Water hammer?
4. What spring rate and damping are optimal in pipe retention members?

H. *Materials Selection and Testing*
1. Can filament winding compounds be specified which have total resistance to effect of fresh water of analyses similar to that of the considered source streams?

2. Can these compounds be shown to be impervious to the action of the sea's microbiology? Even with exposed ends of fibers?
3. What scaling considerations exist in evaluating the validity of accelerated life testing in the structures laboratory?
4. How much can life testing be condensed in terms of time?

I. *Pipeline Design Considerations*
1. What pipe joint configuration will be most effective in surviving expected loads at lowest cost? Are there close alternatives?
2. What is the most effective design of retention members?
3. What arrangement of structure is optimal for canyon crossings while affording maximum reliability and maintainability?
4. How close to each other should two or more pipelines be placed for maximum route efficiency and maintainability?
5. What cleaning techniques are indicated? Will cleaning damage the pipe material? How often would the cleaning have to be done to maintain proper buoyancy? What would the cleaning cost?

J. *Pipe Manufacturing*
1. How are prospective vendors preparing the qualification of their materials, processes, and large-scale production capability for possible bidding on future project contracts?

K. *Definition of Optimal Marine System*
1. What arrangement of vessels and equipment possesses the best cost effectiveness? What are reasonable alternatives?
2. What is an optimal rate of installation versus progress of work?
3. What would be the cost of operating submerged observation and maintenance vehicles?
4. What frequency and sophistication of water quality analysis is indicated to guard against chemical sabotage? What is the annual cost?

SUMMARIZING QUESTIONS
With completion of the study, what will be the reconnaissance level estimate of:

1. relative technical feasibility of any configuration of the undersea aqueduct?
2. economic feasibility of one or more configurations of offshore conduit?
3. optimal arrangement or combination of conduit configurations?
4. the need for a pilot project for larger scale testing? If so, what scope?

A reconnaissance study would probably take approximately four years to complete if it includes the accelerated life testing of medium-scale pipe specimens. Two years will be required for the ocean surveys and subsequent completion of analytical work leading to the specification of desired physical properties of materials to consider in pipeline design. Within six months specimen materials would be procured and the tests could commence.

APPENDIX

Design of Marine System for Installation, Operation, and Maintenance

The complete marine system for the aqueduct project would include at least the following functions:

a.   route survey and data acquisition;
b.   aqueduct system installation;
c.   aqueduct system surveillance;
d.   aqueduct system maintenance; and
e.   aqueduct system repair.

The optimum design will be the one that integrates most completely these functions with a resulting least total life cycle cost.

Following the system approach, complete alternative systems which accomplish all of these functions should be formulated and evaluated against performance and cost.   This requires a comprehensive analysis of all sub-functions and their interrelationships; tabulation of performance criteria, requirements, and constraints; and, the derivation of design requirements.

Once the best total system concept is selected, it is possible to proceed to optimize the various sub-systems with confidence that system optimization will not be compromised. The sub-system design procedure remains the same, i.e. formulation of alternatives, evaluation, and selection.

Recognizing the importance of rigorously adhering to this systems procedure, it is with some hesitancy that a concept for installation and maintenance is included in this report.   It is presented only in the context of exploring the feasibility of the present state-of-the-art to accomplish these tasks by describing one workable concept.

Although the primary objective is to design an offshore aqueduct system, attention must be given to practical considerations attendant to fabricating, handling, and transporting aqueduct sections, mooring devices, etc.   As a result, a preliminary concept has been developed to identify typical types of equipment required to fabricate, assemble, and support the aqueduct.   The concept presented herein includes a vessel and other support equipment necessary for the aqueduct system.

One of the basic criteria for the aqueduct installation and maintenance vessel and associated support equipment is to provide a very stable offshore platform with excellent maneuverability.   The ship should be capable of laying, inspecting, servicing and repairing the aqueduct piping and its associated mooring devices on the ocean bottom in state 4 to 5 seas in water up to 500 feet in depth.

DESCRIPTION OF A SYSTEM AND ITS OPERATION

The concept visualizes use of a Tri-Sec Catamaran (TSC).   A Tri-Sec hull is a new stable hull form under investigation by Litton (see Exhibit 9.2).   It is equipped with revolving cranes and is capable of maintaining an accurate position over the end of a pipeline.

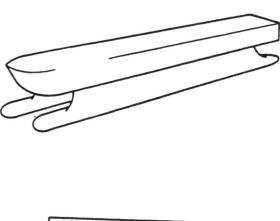

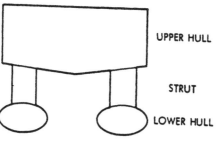

Ex. 9.2 Tri-Sec Hull

Ship positioning might be performed by a main propulsion system of four cycloidal propellers automatically controlled by computers monitoring data from strategically located underwater transponders. The pipe sections and their mooring anchors would be transported to the TSC by barge, by means later to be described, and placed in position on the ocean floor by the TSC (see Exhibit 9.3).

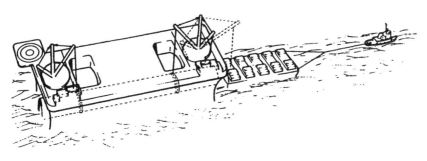

Ex. 9.3 Handling pipe and mooring anchors

The aqueduct piping would be fabricated at a manufacturing facility ashore. Each section of pipe (for the purpose of this discussion, 400 feet in length) would then be provided with sufficient temporary buoyancy to float the pipe in the sea. The temporary flotation devices (see Exhibit 9.4) would be eccentric to float pipe sections in any desired position. The pipe sections would be hoisted onto a barge and transported to the marine site where the sections can be jacked and/or rolled off the

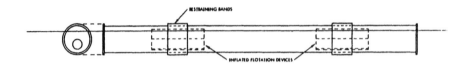

Ex. 9.4 Pipe flotation devices

barge using a parbuckling procedure such as shown in Exhibit 9.5. The temporary flotation devices would be removed and, after final preparation, the pipe would be lowered into position and installed as a section of the undersea aqueduct using an assembly fixture.

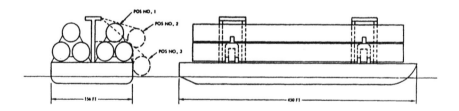

Ex. 9.5 Aqueduct pipe barge

The primary undersea aqueduct pipe-laying ship would probably have a length of from 400 to 500 feet, a width of from 160 to 200 feet, a depth of from 70 to 90 feet, and a draft of from 25 to 30 feet. The ship would be equipped with two revolving cranes, each capable of handling the proposed anchoring devices and together capable of handling a 400-foot section of pipe (approximately 400 tons). The cranes would be located on the main deck at each end of the vessel to service the mooring devices and miscellaneous lifts in operations off the ends of the ship and are able to rotate inboard to handle pipe sections through open wells, as in Exhibit 9.6.

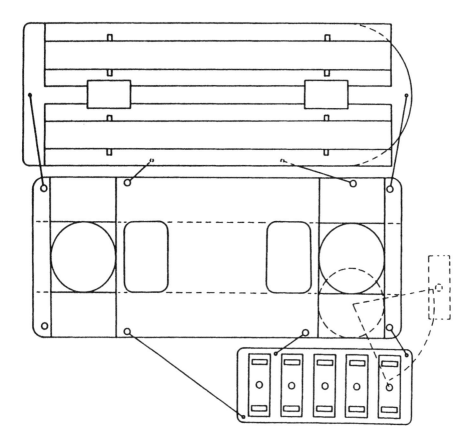

Ex. 9.6 Alternate handling method for pipe and mooring platforms

An alternative approach to crane coverage would have the revolving cranes mounted on rails thwartship on the main deck to provide crane coverage on each side of the ship fore and aft. The TSC might be powered with four diesel or electric-driven cycloidal propellers of some 1200 to 1500 shaft horsepower each. The propellers would be located at each end of each of the twin catamaran hulls to provide protection against operational damage. A propulsion system of this type would permit the ship to maneuver successfully (without encumbering rudders) in any given direction and permit the vessel to remain virtually stationary above the end of the pipeline in any weather otherwise suitable for a pipe-laying operation.

The TSC would be essentially a double-ended vessel capable of travelling at about 12 to 14 knots sustained speed in forward or reverse. A minimum crew would be required to man the TSC through the use of anticipated automated controls and equipment. Accommodations for the operating crews would be provided aboard the TSC, and it is probable that a helicopter landing facility should be provided on deck.

DESCRIPTION OF THE AQUEDUCT LAYING PROCESS

The aqueduct mooring devices could be manufactured ashore and transported to the operating site on barges under tow. The barges could be maneuvered into position forward of the TSC or alongside as in Exhibit 9.7. Lines from the TSC would be secured to the barge and tended by winches on board in the "tow" configuration (refer back to Exhibit 9-6). The tug would maintain tension on the barge and positioning lines from the barge to the TSC. With the barge in position and under positive control, one or two mooring anchors would be hoisted by a revolving crane and set on the TSC main deck outboard of the crane position where they will be fitted with securing hardware and prepared for lowering to the ocean floor. When ready, the mooring devices could be lowered to the ocean floor by the revolving crane or transferred to a deep sea winch for actual lowering to the sea bottom.

An alternative method for transferring pipe sections from alongside the pipe barge at the operating site would be to have a medium-sized tug with a low- profile tow the pipe section into the TSC tunnel. For example, there are several sea-going tugs of this type that are capable of navigating the New York State Barge Canal with maximum overhead clearings of 16 1/2 feet. One of the low- profile tugs could tow the pipe section into the TSC tunnel, transfer control of the pipe to the TSC in the

Ex. 9.7 Aqueduct pipe handling

tunnel, and then proceed through the tunnel emerging at the far end. To adequately control the pipe section during this maneuver, a second tug would be secured to the aft end of the pipe section and would also provide a stopping or backing force for the pipe section. Another alternative may prove to be the best solution to the pipe transfer problem. In this approach, a pipe section would be parbuckled into the water just ahead of the TSC and left floating. The TSC, with its great mobility, would then maneuver into position over the pipe section, secure the pipe with on-board hoisting equipment, then return to her pipe-laying station with the aid of computerized electronic aids.

Exhibit 9.8 shows an alternative scheme for transferring pipe from the barge to the TSC. The barge tugboat keeps lines between the TSC and barge taut while transferring pipe sections to the TSC.

Exhibit 9.9 shows an alternative TSC design having pipe storage holds on each side of a pipe lowering well located in the center of the vessel. Swinging support columns permit transfer of pipe from barge to storage holds and then to the lowering well. An advantage of this design is that it enables the TSC to remain "on station" most of the time at sea without additional maneuvering in pipe transfer.

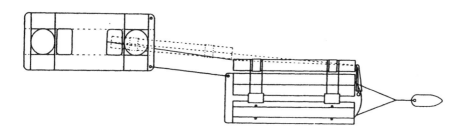

Ex. 9.8 Alternate pipe handling method

Ex. 9.9 Alternate TSC configuration

MAINTENANCE CONSIDERATIONS

Maintenance equipment should be compatible with the total marine system as discussed. Commonality of all equipment is desired and may include sliding bulkheads, strong back fixtures and submerged winching arrangements powered by submarine vehicles. Technical questions concerning practical maintainability include the following:

a. What sea load maxima must be expected during work in process on a minor or major repair procedure?

b. Must the exterior of the pipe be cleaned of sessile animal encrustations? If so, how often?

c. What prospects exist of fouling of the interior of the aqueduct?

d. What aqueduct flow velocity is required to prevent intrusion of sea water into an open pipe?

Regarding question d, analysis based on the relationships of gravity head to sea water intrusion assumes a 30-foot diameter pipe 1.5 x 6 feet long and 260 feet below sea level. This assumption makes provision for the turbulent flow of the emerging fresh water. The complex interaction of the turbulent flow on the boundary layer caused by intrusion is still being studied in the literature.

10

Ocean Farming

Michael Markels, Jr.
President and Chairman, Ocean Farming, Inc.

FARMING THE OCEAN

Ocean farming is a large undertaking, requiring the creation of a new industry on a worldwide scale. It differs from ocean aquaculture in that ocean farming is based on the enhanced production of plant life in the ocean rather than on feeding land-grown food and fish meal to fish. Therefore, ocean farming will increase the total amount of high-quality food available rather than merely increasing the value per pound of product. It also will require some portion of the fisheries industry to change its role from that of hunting and gathering of wild stock to that of participating in a farming operation; from exploiting a "common" resource to investing in the creation of a privately held productive resource.

The fishermen of the world have known for many years that there is great variation in the productivity of different areas of the oceans. Within the last ten years, the extent of this variation has been measured and the reason for it determined. The necessary nutrients to support a phytoplankton bloom only occur in a small fraction of the ocean surface. This gives us a picture of the ocean as a vast desert with only a few verdant zones where life abounds.

It is easy to spot the difference. For most of the ocean, one can see 150 feet through the water just as in the Gulf Stream. In the productive zones, one can see only a few feet because the living matter is so dense. This is the case in the upwellings off the coast of Peru where the difference is obvious. The productive zones are rich in iron, phosphorous, trace metals, silica, and nitrate. Each ocean zone must be sampled and the nutrient requirement ascertained to bring it to the level of the Peruvian upwelling. In the barren tropical oceans, we expect the main missing element will be iron with some phosphate.

It is estimated and now generally accepted that 60% of all the life in the ocean arises from 2% of the ocean surface. Therefore, if all the ocean was like that 2% verdant zone, we would have 0.6/.02 or 30 times the present ocean life. If all the ocean was like the 98% nutrient-poor zone, we would have 0.4/.98 or 0.41 times the present ocean life. The ratio of verdant zones to nutrient-poor zones is therefore 30/0.41, or 73.5 times. That is, if we fertilize a nutrient-poor region of the tropical ocean to conditions such as exist off Peru, we should get an increase in phytoplankton production of 73.5 times.

A recent paper by Pauly and Christensen (*Nature*, March 1995) gives a measure of the "primary production ratio," including catches and discards. This is the pounds of fish caught per pound of phytoplankton produced. The open ocean value is 1.8%, but the tropical upwellings value is 25.1%. This gives a picture of transfer of biological material between trophic levels that is much more efficient than previously thought. Fish farming gives values of pounds of fish produced per pound of feed of 50% to 90%. The key is to be sure that the fish expend minimal energy to obtain their next meal. In order to achieve the 25% value, the efficiency of transfer between trophic levels must be between 50% and 70%. This is only possible in a very dense eco-system where the energy loss for capture is small, as occurs in fish farming.

According to the Pauly paper, the increase in fish catch per 100 pounds of phytoplankton jumps from 1.8 for the open barren ocean to 25.1 for the tropical upwellings, a multiple of 14 from extra nutrients. Multiplying this increase by the increase in phytoplankton gives 14 x 73.5, or 1025.

Some confirmation of these trends can be obtained from data on the effects of the El Niño event of 1982-83 (and will undoubtedly prove to be true from the 1997-98 El Niño as well). The anchovetta catch was reduced to 1/600th of its normal value. Since the fishing effort per ton of catch went up during the event, we expect that the fish stock went down by a factor of about 1,000. This gives a reasonable check with the factor of 1,000:1 estimated above. It is interesting to note that such large changes in productivity at all levels of the food chain took place in a time frame of a year or so, indicating the likelihood of a similarly rapid response to ocean farming in the tropical ocean.

OCEAN FARMING PARAMETERS

The ocean differs from the land in several respects: (1) there is never a drought, (2) it moves, and (3) it mixes both vertically and horizontally. The first difference means that we only have to add minor constituents. The second difference means that where we add nutrients and where we harvest are likely to be many miles apart depending on the current. The third difference means that we must do our farming in the open ocean on a large scale or we will never be able to find the results. Finally, we must do our fertilization in deep water so that the deep ocean currents can process the rain of organic materials produced without becoming anoxic, leading to conditions that will kill the very fish we wish to produce.

What are the parameters of ocean farming? First and foremost, it must be done on a large scale relative to farming the land because of the movement and mixing of the ocean surface. There is an optimum size where the edge-to-area ratio becomes so small that the fish are essentially trapped within it. All except migratory fish tend to remain in the verdant waters.

Second, it differs from aquaculture in that it is based on the enhanced production of plant life in the ocean waters. The Redfield ratios describe the response of ocean plant life to critical nutrients. One pound of available iron can lead to the production of 100,000 pounds of biomass. To the iron we may add some phosphate, a float material to keep the fertilizer in the photic zone and perhaps a binder that releases the fertilizer slowly into the ocean water. By the time we have done all this, one pound of fertilizer produces about 10,000 pounds of biomass. The ocean is not a controlled, uniform resource. Therefore, we conservatively estimate that one pound of fertilizer will produce 4,000 pounds of biomass in barren tropical waters.

The productivity per acre should be higher in a nutrient-rich ocean than on land. However, we use 40 tons per acre per year, which is the same as for sugar cane cultivation. That calculates to be 25,600 tons per square mile per year.

The one pound of fertilizer should produce about 200 pounds of catchable fish. We expect these fish to be like the fish caught off Peru, including both the filter feeders and the higher trophic-level fish such as tuna and swordfish.

We will have some control over the fish produced since we will seed the ocean with the fish we want and catch the ones that give us the greatest return for our efforts. We plan to spread about 5 tons of fertilizer per square mile per year and harvest about 1,000 tons of fish per square mile per year.

When dealing with land farming, we are familiar with planting and fertilizing in the spring and harvesting in the fall. In the ocean, under ideal conditions, phytoplankton doubles every day or two, producing a bloom of 20 to 30 times in about five days, seven hundred to one thousand times in ten days. Then the zooplankton graze on the phytoplankton, the bait fish eat the zooplankton, and on up the food chain to the large mammals and apex predator fish whose life cycles approach decades. We plan to fertilize in areas of the open ocean where the currents maintain the fertilized water within our control for at least twenty days, consistent with the life cycle of the upwelling-fed blooms off Peru. Longer available time for the blooms will reduce seeding requirements for both plants and fish, and therefore increase productivity of the resource.

The credibility of these predictions has been greatly enhanced by the publication of the results of the IronEx II experiments in the October 10, 1996 issue of *Nature Magazine*. In these experiment ferrous sulfate was added to the waters of the tropical Pacific ocean in an area of high nitrate, low chlorophyll, HNLC, water. Exhibit 10.1 shows variations in chlorophyll, nitrate, CO_2 and iron over 17 days as the result of the first iron addition (day 0), the second iron addition (day 3), and the third iron addition (day 7). The chlorophyll bloomed on days 5,6 and 9. The chlorophyll concentration increased by a factor of twenty-seven times by day 9 in spite of a loss of about 95% of the iron to precipitation. We expect to achieve essentially 100% utilization of the iron by phytoplankton growth in our fertilizer system. These results are the first to show that iron is the controlling nutrient in these high nitrate-low chlorophyll open ocean waters.

Ocean farming operations may be carried out in ocean waters that are also low in nitrate by fertilizing with the other missing nutrients, principally iron and phosphorous, causing a bloom of nitrogen fixers. This happens in the open ocean where dust containing iron fertilizes the ocean, creating large fisheries such as those off Alaska and the northwest coast of the U.S.

Ocean farming may affect our concerns about the atmospheric CO_2 balance and global warming. The U.S. burns about 1,390 million tons of carbon per year in the form of fossil fuels (gas, coal, and oil). With 3.67 tons of CO_2 produced for each ton of carbon burned, the U.S. produces about 5.1 billion tons of CO_2 annually. One ton of fertilizer produces 4,000 tons of biomass and removes (initially) 5,500 tons of CO_2 from the ocean. About half of the biological material will descend to the ocean bottom in the form of droppings, plant matter, and shell carbonates, where it will ultimately be picked up by the bottom currents and eventually recycled into upwelled water on a geological time scale. Therefore, to sequester to the deep ocean the U.S. CO_2 production from the burning of fossil fuels we need to spread about 1,500,000 tons of fertilizer. This amount of CO_2 can be taken out of the ocean and the atmosphere by fertilizing an area about 300 miles wide and 1,000 miles long. The total carbon that becomes part of this cycle is thus removed from the ocean waters and the atmosphere for one to two thousand years, giving an avenue of positive action to ameliorate our concern with regard to the effects of burning fossil fuels on the world climate. This sequestering in the deep ocean can be done for about $3.00 per ton of carbon sequestered.

While we do not now know all of the environmental impacts of converting areas of the ocean from barren deserts to verdant blooms we can outline some of the expected effects. Since plant life will be dense, fish will expend less energy to get to their next meal, and the ratio of pounds of fish per pound of phytoplankton will increase greatly.

Whales and porpoises will increase in the fertilized area, gaining weight rapidly during the time they are there. These are migratory species they typically congregate where the food supply is plentiful. Over a long period of time the total world count of porpoises and whales will increase slowly due to the long doubling time for these species. Such a positive trend as the result of ocean fertilization could, of course, be reversed by adverse actions in other parts of their habitat. The effect on large pelagic and migratory fish will be similar. Tuna, for example, increase rapidly in body mass during the time they are in verdant waters. Then they move to breeding grounds where they spawn. The increased food availability will increase the numbers of tuna, bill fish, and dolphin in the fertilized area as they seek new food sources. They will be very happy fish.

There are other ecosystems in the fertilized area that may not be as happy. Coral reefs have survived by evolving their ability to grow in low-nutrient ocean waters. When nutrient levels are increased, algae grow faster until a level is reached where the nutrients produce so much algae that it shades the coral and kills it. It may be that some coral must be shaded in order to achieve the increased productivity that we seek. In commercializing ocean farming, large areas of the tropical ocean will be involved. Therefore, some adverse effects on local corals could occur. Efforts should be made to minimize any such adverse effects by avoiding coral.

The great environmental plus for ocean farming is that, unlike erosion on land, none of the changes are permanent. We only have to stop fertilizing and all traces of the nutrients are gone in less than one month.

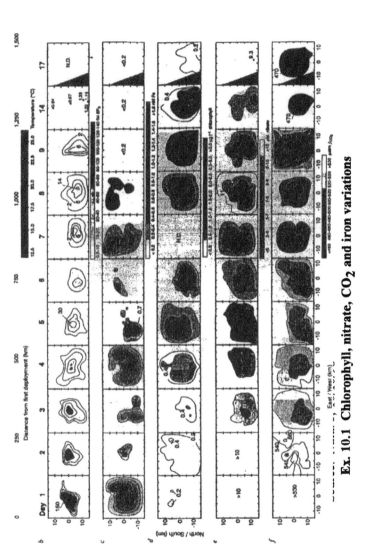

Ex. 10.1 Chlorophyll, nitrate, CO₂ and iron variations

The overall effect of ocean farming will be to increase greatly the amount and diversity of the marine ecosystem in the fertilized zone. This is a positive answer to the worldwide problem of over-fishing, since we will always create more fish than we harvest. This will be done in the context of private property rights so that conservation and the creation of value will be a part of everything that is done. We can also answer concerns regarding global climate change through CO_2 sequestering.

TECHNOLOGY DEVELOPMENT

The proposed fertilizer materials will have special features, such as particle size, dissolution rate, density, and ratios of critical nutrient constituents. Since sea life appears able to process nutrient materials regardless of chemical makeup or form as long as it remains in the photic zone, we believe that the least expensive, most readily assimilable forms of raw materials having the appropriate chemical compositions should suffice. The fertilizer must not contain traces of toxic chemicals, especially those known to bioaccumulate in marine organisms as they move up the food chain, and they must also be free of pathogens that could be passed up and ingested by fish destined for human consumption.

The concept of fertilizer design is to obtain a rapid phytoplankton bloom that fixes nitrogen and further promotes accelerated growth of oceanic biomass at successively higher trophic levels. To do this, the buoyant fertilizer system should contain the limiting nutrients such as iron, phosphate and other trace nutrients. A strain of phytoplankton specifically selected to initiate the process may also be seeded in the broadcast stream. There are difficult technical problems associated with the design of the fertilizer. The added nutrients must be in a form that permits them to remain in the ocean surface water for an extended period and not sink to the bottom as a precipitate. The optimum ratio of phosphate to iron and any other missing nutrient must be determined in order to design the fertilizer system for the ocean area selected.

EXPERIMENTAL PROGRAM

A three-phase technology demonstration program has been designed and is presently underway.

Phase I. Fertilizer Development

In this phase ocean fertilizer materials will be designed to assure that they meet the requirements of density, solution rates, and performance. The ability of the fertilizers to support a phytoplankton bloom under laboratory conditions is tested. This phase is now essentially complete.

Phase II. Fertilizer Evaluation and Refinement

This phase tests the phytoplankton response to fertilizers developed in Phase I in open barren tropical ocean. A voyage was carried out in the Gulf of Mexico in early January 1998. Three 9-square-mile patches were fertilized: one with iron only; one with iron and 6.35 times the molar ratio of phosphorous to iron; and one with iron and 63.5 times the molar ratio of phosphorous to iron. The iron was in the form of a chelate to protect it from precipitation; the phosphorous was in the form of phosphoric acid. The ocean and weather conditions, including a deep thermocline and high winds, caused the fertilizer to mix much more rapidly, both vertically and horizontally, than planned.

The result was a bloom of large diatoms to 4.3 times their initial concentration in about one day. After that, the mixing diluted the signal to about 1.5 times the initial chlorophyll concentration. These results, while giving a positive indication of a large bloom, are not definitive and do not provide a verifiable measure of phytoplankton increase over the period of the expected bloom of about two weeks.

A program of six additional tests is planned, beginning in May 1998, to obtain the definitive results sought.

Phase III. Full-Scale Fertilized Ocean Testing
This phase will demonstrate the production of fish from fertilization of barren tropical ocean. It will require the fertilization of a larger area than Phase II and for a longer time. The fertilized area will be seeded with filter-feeder fish that live on the phytoplankton produced and their growth rate determined. The fertilized area will be about 500 square miles, depending on the currents and mixing of the ocean surface, and will be away from coral reefs. The test will last about six months and will probably be done in the vicinity of the Marshall Islands where both high and low nitrate waters are available.

Experienced organizations are already under contract to carry out the three-phase program. With the successful completion of these experiments, the fertilization of the ocean will be demonstrated and the resulting increase in fish production documented. Commercialization of ocean farming can then begin.

COMMERCIALIZATION
Ocean Farming, Inc. (OFI) is a new company organized to develop and commercialize a revolutionary patented technology to increase the productivity of the open ocean, raise and harvest fish stocks, provide needed protein to feed the world's expanding population, and sequester CO_2 in the deep ocean for one to two thousand years.

In order for commercialization to proceed, OFI must have a demonstrated fertilizer, as outlined above. It must also have a suitable platform to carry out the venture. This requires: a large tropical ocean area; deep, barren ocean water; benign currents and weather; and most of all, private property rights for the purpose of ocean farming.

The large area is required because we will not physically fence off the fertilized area, but will rely on the size of the verdant area to keep the seeded filter-feeder fish there. The higher trophic-level fish should also remain near their source of food as they do in the upwellings off Peru. We need tropical waters in order to have constant sunlight and deep water so that the biological material that sinks to the bottom will not become anoxic. We need barren water so that it will contain the fish and our fertilization will have the largest effect. We need benign currents and weather so that the fertilization effect remains within our control for the required twenty days.

The key requirement is private property rights. We cannot invest in making the ocean more productive and then have others reap the benefits. After considerable effort on both sides, we have signed an agreement with the Republic of the Marshall Islands (RMI) granting OFI private property rights for ocean farming in part or all of their 800,000 square mile economic exclusion zone. We have selected a 100,000 square mile segment as our first commercialization area. This should be more than adequate for some years to come since, at 1,000 tons per square mile per year, the 100,000 square miles should produce 100,000,000 tons of catchable fish, essentially equal to the current world production. We expect that it will be many years before we reach this level of production. How rapidly we ramp up will depend on the investment required and the demand for fish, worldwide.

The fertilization of 100,000 square miles of tropical ocean in the RMI should sequester about 30% of the U.S. CO_2 production, resulting in income for OFI from CO_2 emission credits and reduction of the concerns regarding global warming.

EXPECTED RESULTS
The average price of fish "at the dock" is about $0.40 per pound worldwide. The cost of providing that fish is somewhat more than that, the difference made up by government subsidies. Of the various components of that cost, finding the fish is the largest single item. We do not expect to have a high cost in this category since we will have put the fish there and will manage the fishery to ensure a high density of plant and fish life. We expect that the cost of fertilization will be low, below $0.01 per pound of fish. The total cost of producing the fish and catching them should be about $0.05 per pound. The cost of processing $0.05 per pound and the cost of distribution $0.05 per pound, giving a total cost of $0.15 to $0.20 per pound after all overheads are included. It should be clear that these are only best guesses at this time. The main point is that there is considerable room for profit with an average price at the dock of $0.30 to $0.40 per pound for the product.

The sale of CO_2 emission credits from sequestering in the deep ocean will have to await their inclusion in the U.S. government's global warming initiative.

SCHEDULE
We expect to complete the technology, demonstration and testing program, Phase II, in 1998 and Phase III in 1999. Work in the RMI should begin in 1999, with the required testing of the waters for subsequent fertilization testing. This would allow us to start commercialization in the year 2000. We will start small, perhaps less than 10,000 square miles. This has considerable risk since we are sure that the project will be successful if the fertilized area is big enough, but we do not know how small it can be and still succeed.

THE ENTERPRISE
The total enterprise required to make ocean farming a reality is comprised of many individual parts. Corporations exist that have great expertise and experience in carrying out their specialties. We therefore do not plan to replicate these capabilities within OFI but rather to form a team, with contracts specifying the relationships required.

We expect OFI to manage the enterprise, carry out the research and development required, contract for services, own patent rights, and maintain private property rights for the ocean area in the RMI and other suitable ocean areas.

OFI will contract for the production of the patented fertilizer, the transportation to the RMI and the spreading of the fertilizer on the ocean surface, all to OFI's specification. We have a large fertilizer producer who has agreed to carry out these tasks.

We will also contract with others to seed the fertilized area with desirable fish, harvest the fish produced, process the catch, and transport it to market. These tasks will also be done to OFI's specification. We will be careful always to produce more fish than we harvest from the ocean. We also plan to have no discards or fish waste. We should be operating at a scale that will make it possible to utilize everything we catch. This should also increase the profitability of the enterprise.

This is a new concept, based on new technology. There is much to be learned as we apply it to the ever-changing ocean. It is clear that the return to mankind from the success of this endeavor leading to the farming of selected portions of the almost three-quarters of the earth covered by the oceans, will be great indeed.

Section Three

COMMUNICATIONS AND TRANSPORT

11

World-Scale Seamless Transportation

Ernst G. Frankel
Professor of Ocean Engineering, M.I.T.

A major part of the origin-destination (O-D) cost and time in intermodal transport traditionally has been consumed by intermodal handling, storage, and other transactions. In a typical bulk O-D transport between the midwestern U.S. and Continental Europe in 1980, only 22% of the time and 31% of the costs were consumed by actual transportation. While container shipping of similar cargo on the same route improved the situation, the 1980 study found that actual transport time and costs in that market were still only 44% and 51%, on average, respectively, even using full container O-D transport. It is increasingly recognized, therefore, that major time and cost improvements in O-D intermodal transport are most effectively obtained not by increasing the speed and reducing the cost of transport, but by reducing intermodal transfer times and costs.

New technology, infrastructure, and most importantly management techniques now allow close modal delivery time control with great reliability. This use of the just-in-time (JIT) concept as a way to schedule time in turn permits the development and effective use of direct intermodal transfers by using new intermodal technology with little if any intermodal storage or buffering. Such seamless transfers are furthermore facilitated by new information and communications systems that encourage paperless intermodal transfer management while assuring real-time information management.

Intermodal transportation has become an integrated system of coordinated activities and services which move cargo from origin to destination most expeditiously and inexpensively. The revolution in supply chain management and the increasing concern with the elimination of intermodal cost and time are leading us toward seamless intermodal systems in which intermodal transfers are as continuous as possible. Modal operations are closely matched with capacity and schedule to assure that modal carriers can transfer much of their cargo by direct inter-carrier transfers without intermediate storage. This also requires the matching of cargo and empty flows and integrated cargo flow and carrier movement control in the whole intermodal network.

The issues that must be addressed and the problems to be solved that will lead us toward such seamless intermodal transport are discussed in this chapter. The roles of various technological developments which permit seamless intermodal transportation are reviewed and the future of such systems are projected.

DISINTEGRATED INTERMODAL TRANSPORT

Although we have long used the term "integrated intermodal transport", intermodal transport has been highly disintegrated most of the time. Intermodal operators such as Sealand, which not only introduced container shipping but also insisted from the start on owning/controlling the various modes and intermodal terminals, was for a long time an exception. Others essentially operated as members in multi-modal chains, with transport links and intermodal terminals managed as individual or separate operations, each with its own rules and objectives. Modal transport or terminal provider usually optimized their individual operation. This sub-optimization of individual links often caused a less-than-efficient total intermodal operation.

Link operators have different and often conflicting objectives which not only affect the matching of links in terms of capacity but also in terms of operational decisions. For example, ports want to maximize the use of their equipment and facilities and may therefore encourage the use of stacking/storage facilities and equipment. The lack of effective integration of modal links and intermodal terminals often causes gross under-utilization of capacity or major delays. As each link or terminal attempted to maximize its profits, decisions are often made which add cost and time to other elements in the O-D chain.

Another result of this approach was larger-than-necessary fluctuations in demand for and use of link or terminal capacity and resulting unnecessary overinvestment. In the last few years, shippers and consignees have become increasingly aware of the added cost of waiting, low capacity utilization, and double handling in terms of money and time. Effective supply chain management has become a serious challenge which is being addressed aggressively worldwide. It demands reduction and/or elimination of intermodal waiting time and double handling.

A major and rapidly increasing force driving demands for seamless total quality supply chain intermodal transport is outsourcing. More and more producers in services, manufacturing, and materials are sub-contracting major functions to various suppliers in order to improve operational and scale efficiencies. This is a boom to integrated intermodal transport on one hand, but it requires a complete change in its structure and management. Intermodal gaps must be closed, just-in-time must become a fact and not just a slogan, and seamlessness requirements must be continuously updated.

Disintegrated intermodal transport, in which each modal transport or intermodal terminal link tries to optimize its operations by investment, and operational management which tries to maximize that link's profits, was bound to fail users that were interested in total multi-modal origin-destination times and costs. The performance of a multi-modal system in which each transport and terminal suboptimizes will always be far from optimum. It has long been known that the sum of the suboptima of the links is rarely the optima of the sum of the links or the system of link networks.

It has been recognized that the disintegrated management of intermodal transport links was the major factor influencing time and cost performance of intermodal or multi-modal transport systems. Not only did it result in overinvestment in and mismatch of transport and terminal links, but it also caused major operational problems and resulting waste. Some links, like mainline shipping, are highly intermittent, often with only one large unit per week that demands fast turnaround; other links, like a container port terminal, prefer near continuous flow operations. As a result, compromises have to be made. For example, ports need to provide reasonable ship/shore transfer rates which satisfy the liner operator without penalizing the port by under-utilization. Obviously, large ports with many calls from nearly identical-sized ships, could benefit by scheduling ship arrival/berth assignments so that more ships share the same efficient berths.

It has also been recognized that to improve port utilization, ship and inland transport turnaround performance requires close cooperation, compromises in scheduling and routing, total quality management, and an open communicating system in which each party maintains continuous contact with and adjusts the management requirements of all the other parties affected at the port interface.

In summary, intermodal transport of containerized and other goods is still highly disintegrated, a fact which reduces many of the advantages of intermodal transport.

APPROACHES TO REDUCE INTERMODAL SEAMS

There are various possible non-mutually exclusive approaches to the reduction of intermodal seams. One is the increase in service frequency to make it near continuous. As one ship leaves, another arrives; trucks and other land transport operate in a virtually constant stream consistent with the loading/unloading rate of the ships. This way cargo could theoretically be transferred directly between ships and trucks without the use of intermediate storage.

This concept underlies the trend toward formation of large shipping alliances that are able to increase significantly the frequency of ship calls and the size of the container trailer/truck fleet.

While the last twenty years have been the age of the shipping consortia, we are now building toward the age of shipping and intermodal alliances, particularly in liner shipping. The incentive for this is the increasing demand for intermodal service integration, in which several major operators not only share space on each other's ships but also share inland depots, feeders, container terminals, and in future, container inventories. As ships become larger and 5000-6800 TEU vessels (post-Panamax) become a reality, with increases in the relative capacities of land and ocean carriers, it is increasingly evident that improvements in costs and operating efficiencies will in future have to be extracted from better terminal interfaces, utilization of inland and feeder operations, and improvements in the use of container inventories -- particularly the relocation of empty containers, which is still one of the largest costs to operators. Most importantly, the seams in terms of cost and time spent must be eliminated between modal links. This may result in the designation of universal containers which can be readily exchanged among alliance members without loss of advertising identity.

The concept of global shipping alliances is shifting toward global intermodal alliances which encourage demand for the seamless integration of modal links worldwide and on a large scale. To eliminate seams in intermodal transport requires large scales of operations and related frequent ship and land transport services. These types of grand alliances have been formed in the Atlantic, Pacific, and on other long distance deep ocean trades.

Managing such grand intermodal alliances requires effective integration of

- large scale of operations,
- large unit sizes of ships,
- integrated modal services,
- combined feeder services,
- consolidated container pools and inland depots,
- information and communications systems,
- marketing bases,

as well as new approaches to reduce or eliminate intermodal seams in physical and information transfer. Most importantly, we must aim at reducing intermodal physical and information storage, delays, and handling.

To that end such alliances must go much further than just integrating their operations into combined services. Seamlessness requires real time and technology coordination of and cooperation in all aspects of operations, as illustrated in Exhibit 11.1. The major objective of seamlessness is continuity in the coordinated flow of cargo and information from origin to destination and effective capacity balance over time between all links in the intermodal chain. As noted, this is becoming increasingly difficult as the size of mainline containerships continues to grow in relation to the size of feeder and land transport. As a result, increasing ship size will drive even larger alliances to assure more effective utilization of terminal links and improved seamlessness in transfer to and from feeder and land transport modes.

MARKET EXPANSION OF INTERMODAL TRANSPORT

Intermodal transport has expanded greatly in recent years and now serves most areas of the world. New high-speed containership routes have been introduced between Asia/Pacific and Europe as well as east coast America. Major new rail links are being planned to connect Asia and Europe. These new rail links are not only designed to improve the linkage between the continents, but also provide new efficient intra-Asia linkages. Although some shippers are skeptical about the economics of trans-Asian rail links, the long-term advantages of such a rail network in opening up Asia -- not just connecting its major markets with Europe -- are obvious.

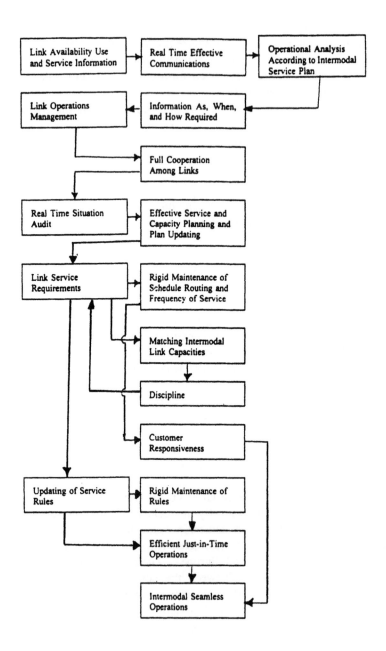

Ex. 11.1 Requirements for intermodal seamlessness

Structural and economic changes in Russia, China, India, and South East Asia are increasingly influencing the type of goods trades, and containerization of intra- and inter-regional trade. Similarly nations and regional groups now recognize the importance of access to the global market place and greater modal transport development coordination. There is increasing recognition that intermodal corridors, particularly land transport networks, must connect efficient intermodal transfer, packaging, and consolidation depots. All of this requires structural changes in intermodal transport systems with large-scale coordination of operations, inventory, capacity, and information management.

TECHNOLOGY DEVELOPMENTS TOWARD SEAMLESS INTERFACES

In line with the increase in the size of containerships, feeder vessels, and unit trains, major developments in interface technology have taken place that are designed to reduce terminal turnaround time of these vehicles and improve the continuity of flow of cargo in the intermodal transport chain. Developments in cargo/container handling transfer and storage technology fall into a number of major categories. In container ship/shore transfer equipment, we now have:

(1) smart spreaders which identify containers, receive/transmit signals, and automatically adjust with respect to changes in container load centers,
(2) effective anti-sway/sway prevention/control technology,
(3) automated pointing control,
(4) single and double trolley cranes,
(5) automated combined horizontal and vertical spreader motion to reduce cycle time and pointing accuracy,
(6) block container handling with telescoping spreader multiple container locking arms,
(7) super LUFF container pallet systems for simultaneous handling of blocks of (empty) containers,
(8) extended backreach container gantries for backreach storage (overstow or prestow) and direct feeder (ship/barge) transfer, and
(9) super post-Panamax with high-speed double trolleys and larger gauge.

At the same time, integrated container transfer and storage/stacking/freight station silos are gaining attention.

Steel-frame container silos can be equipped with container freight stations (CFS) on the ground level and their hoists or elevators used to transfer containers to a space between the import CFS under one tower and the export CFS under another tower. A more advanced approach is to locate the silo towers close to the pier bulkhead and use a double trolley crane to serve both the ship loading and unloading as well as the silo container storage operations. The concept of high-rise container silos integrated into shipside transfer operations is particularly attractive where container terminals are built on artificial islands, reclaimed land, or serve largely as transshipment terminals. Obviously, ships have to be moved alongside the berth to align the stationary cranes with the onboard container stacks. Such ship winching devices have been developed and can move even large ships to new stack rows within a few minutes.

Another interesting development is the floating or barge-mounted container terminal or container stack. The increasing use of feeder container service between smaller and load center ports demands more attention to the effectiveness of transshipment which usually involves double handling of containers (including stacking, etc.). Integrated tug-barge technology is being used to drop off prestacked container barges which serve as floating container stacks for direct transfer of containers to and from mainline container vessels.

The container barges may be equipped with onboard container gantries, served by extra length outreach shore-mounted gantries or by gantries mounted on a special crane barge placed between the container barge and the container vessel.

High-rise container storage systems have been developed by Sealand which constructed a vertical multi-level container storage building in Hong Kong. Several continuous chain parking garage-type container storage systems have also been designed. These assure perfect selectivity, but usually have a large number of working parts. Also their cycle time is often inadequate. The major objectives of these developments are:

- an increase in container handling rate to 30-50 boxes per hour per gantry, and 90-150 boxes per hour per berth,
- greater stack selectivity to reduce overstow handling costs,
- improved intermodal interface,
- more efficient use of terminal waterfront and storage area,
- ability to handle outsize containers, and
- decrease in ship turnaround time and improvements in intermodal transfer.

These developments are also designed to reduce or eliminate shipside storage/transfer transport cost and time, and assure direct ship/feeder land transport or ship/storage transfer. In many cases, where the berth face is some distance from the stacks and inland transport (rail or road) transfer facilities, transport must be provided. This is most commonly handled by tractor trailers, tractor trailer trains, automated guided transfer vehicles, or container conveyors. Automatically locking and positioning trailer trains (often under the control of an imbedded winching/positioning system) can provide a nearly continuous transfer system. Increased transshipment at major hub ports encourages use of floating relocatable barge-mounted container stacks which can also serve as barge transporters.

Advances in yard cranes with AC drives, wide spans, great heights, anti-sway, auto-pointing and smart spreaders now permit high-speed stacking/ unstacking. They are often integrated with automated trailer train positioning systems.

Important developments toward more seamless intermodal transport are also taking place in information, communication, tracking, identification, and computation technology. Electronic data interchange (EDI) is the direct computer-to-computer communication of business transaction information. The objectives of EDI are to improve accuracy, timeliness, and efficiency of transactions by avoiding re-keying errors, mail delays, and mailroom activities. Automated EDI has been developed into an effective worldwide system using interconnected teleports to provide real-time data transfer and use.

Satellite communications have become a commonly used method for long distance information transmittal and provide efficient real-time transfer of huge amounts of information using modern data compression techniques.

Operators have found that investment in information and computer technology offers a large return in efficiency and productivity improvements over investment in transport/transfer hardware, as most hardware inefficiencies can be traced to information and communications systems defects.

Important developments toward seamless intermodal transport have been introduced by satellite tracking, remote monitoring, and control technology. One-way tracking (from transmitter to satellite) requires a human link in order to take action, which is inefficient. In the future two-way communication will be used not just to track vehicles and containers, but also to monitor and correct settings. At terminals, intermodal auto equipment identification (AEI) systems with AEI readers will be used. While some identification technologies today only read tags at specific locations, remote reading/recording will become standard for movement control in the future. Some systems already combine satellite and terrestrial tracking systems. These new tracking systems will soon permit direct access to data on the Internet by logging in to check on container location and status.

For gate control, video-based container code recognition (CCR) systems combined with optical character recognition (OCR) systems can be used for damage recognition and control and gate passage management.

There are many other technological developments toward more seamless intermodal transport such as direct continuous mechanical rail car/track container transfer systems. The most important, though, are the new integrated intermodal transport operations management systems that develop and continuously update schedules, routings, assignments, and speeds of all the links to the intermodal system.

MANAGEMENT TOWARD SEAMLESS INTERMODAL TRANSPORT

Rapid changes in the structure, technology, network, and markets of integrated intermodal systems require the use of flexible, adaptable, and innovative integrated management. Such management must give top priority to managing technological change. Similarly, it should emphasize total systems and not modal efficiency, in terms of origin-destination transport times and costs. Adjustments must be made continuously to reduce costs through improvements in assigned capacity and fine-tuning of unit size, speed, service, routing, and frequencies to improve utilization while minimizing origin-destination times. This means moving from operational to cargo flow management, which is given priority over modal and intermodal link operational management considerations.

In other words, we give increasing emphasis to reducing the cost and time of the intermodal interface by moving relentlessly toward its elimination. In practice, this may lead to more and more direct mainline to feeder or land transport transfer without intermediate storage; similarly for feeder-to-land or land-to-land transfer operations. The information, communications, and clearance and documentation obstacles have been largely removed, and we must now concentrate on just-in-time control and operational capacity matching of modal links in intermodal operations.

Effective intermodal operational matching models exist and can be effectively used to improve existing systems.

CONCLUSION

Seamless intermodal systems, while not perfectly achievable, are approachable now. The emergence of large global mainline alliances and increasing incorporation of feeder and land modes into these alliances now permits effective matching and systems control toward reducing seams in modal interfaces. Some large global alliances have reduced their average origin-destination time to actual transport time by 20% or more in time, and 12% in cost.

But much remains to be done. The tools are at hand. Most global alliances now formed are long-term arrangements of ten years duration or more, the type of commitment which encourages determined efforts toward seamlessness in intermodal transportation.

12

An Asia-Europe Railway and a Middle East Island

Hideo Matsuoka
Research Center for Advanced Science and Technology,
University of Tokyo

GETTING STARTED

After World War II, rapid economic growth in the developed countries created an affluent lifestyle that is considered "modern," accompanied by liberalism, egalitarianism, and philanthropism. It also brought mass consumption of natural resources and energy, causing long-term degradation of the environment on a global scale.

There are problems also in the developing countries: extremely unequal distribution of the natural resources and energy that are indispensable for economic development, and a rapid increase in population -- the population explosion -- as a result of which world population has doubled in the last fifty years (see Exhibit 12.1).

Some developing countries are adopting policies to achieve economic growth and an affluent life-style, and they seem likely to follow much the same path that the developed countries took. If they do, it is inevitable that the natural and biological environments will suffer even more damage for many years to come.

For the next generation to succeed the present generation, continued consumption of natural resources and energy are necessary, and if consumption

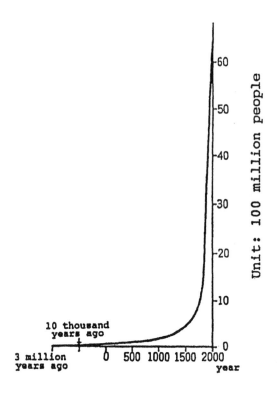

Ex. 12.1 Increase in world population

is constrained, future generations will not be viable. Environmental issues must be considered on a long-term as well as a global basis. What counter-measures macro-engineering can provide against global environmental problems is thus a central question.

MIGRATION

Against the background of the population explosion, large-scale migration of peoples is currently occurring worldwide, in the direction shown in Exhibit 12.2. Regardless of whether or not the migrants are called "economic refugees", they are seeking a new permanent residence to improve their living standards, although some of them may be escaping from various social disorders, including wars and regional disputes.

"From south to north, from east to west," some people aim to go to Western Europe and some others aim at Japan. Migration to the United States is traditional. Do the countries to which they migrate provide attraction to modernism or is modernism the attraction? They migrate to those developed countries where modern technologies are mature and a modern constitution is established. Large-scale migration seeking modern civilization comes from developing countries aiming at modernization, and the migrants are those who opt to move to already modernized countries rather than contribute to modernizing their own countries.

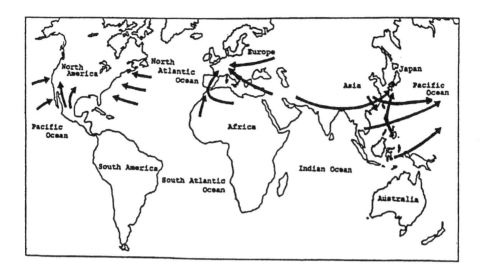

Ex. 12.2 Present-day human migration

Described as "south of the border," they have something of a romantic image, but the problems that the developing countries (which now include the "east" in addition to the "south")are now facing are becoming exceedingly severe on a global scale. Drawn forward by the attraction of modernization, developing countries with gigantic populations are about to progress toward the level of developed countries seeking the affluent lifestyle.

Faced with global environmental problems, the developed countries will have to take domestic counter-measures to protect the environment, but these will not solve the issues. The developing countries want to give priority to economic growth, and it is inevitable that they will give low priority to investment in environmental counter-measures.

A MACRO-ENGINEERING DIPLOMACY FOR JAPAN
In considering the political role to be fulfilled by Japan in international society by changing its hitherto passive diplomacy to a more active diplomacy, I want to propose "dynamic" macro-engineering as a system and technique, in contrast to "static" or traditional macro-engineering.

In the light of the progressive development from operations research (OR), through systems engineering (SE) to macro-engineering(ME), macro-engineering should have superior methodology and practicality. It should not be a genteel discipline to be argued over by library-oriented academics.

Macro-engineering is needed to realize international large-scale public projects, or to deal with global conditions. Needless to say, for the realization of international large-scale public projects, it is essential to solve related international political problems.   What would be a macro-engineering project that aims at resolution of international political problems through realizing large-scale public projects? Thus, by comparison with earlier practice, the purpose and the means are reversed.

Japan has now become an economic giant and can provide substantial overseas assistance every year. Japan has strong human and material foundations and is urged to share international responsibilities in view of its economic power. Then we must face the question, what role and duties should Japan fulfill in international society? And what functions would macro-engineering be able to carry out?

With respect to international large-scale public projects such as the development of outer space, international straits tunnels, a mass and ultra-high speed international air transportation network, transcontinental mass and high speed surface and underground transportation systems, if all requirements for implementation of such projects and funding plans are studied in detail, and a green light is given by a feasibility study which lays a track to lead the project to success, the project will run on that rail without stopping. If a reasonable and firm track is laid as a result of a feasibility study, fund raising will proceed smoothly with social support, and the project will arrive successfully at its terminal station. However, it should be said that such a process assumes a static form of programming.

By contrast, in the case of implementation of counter-measures for international problems or diplomatic policies, the situation will not always proceed as initially planned. Indeed, there are many cases in which a schedule cannot even be established. Any situation unexpectedly occurring or changing would require a dynamic approach for such countermeasures. Countermeasures for global environmental problems must be taken on a global basis, and will be international cross-border projects. The concept of building gigantic systems must not be feared because of their large scale, whatever the late E.F. Schumacher might say. As for the method of coping with these, both symptomatic treatment and causal treatment, including even partial emergency evacuation, will need to be considered.

Almost all of the population increase is now occurring in developing countries. The developed countries supported the rapid increase in population in the developing countries because of their traditional human rights doctrine. But if we assume that the earth is limited, this doctrine might be fundamentally wrong. This is stated not qualitatively but quantitatively. The global environmental problems must be solved as diplomatic problems, placing natural resource and energy problems and the population explosion problems within an achievable range.

With respect to counter-measures for the population problem, particularly stabilization of the population explosion, eventually these cannot help relying on modernization of the developing countries. For that, time is needed, and naturally an increase in consumption of energy will take place. During the time necessary for the stabilization of population through modernization, it will be unavoidable to promote economic growth as a symptomatic treatment on a global scale to deal with the transition state.

At present when the postwar structure (control over the world by the dual American/Soviet hegemony established after World War II) has collapsed, international politics concerning global environmental problems have been highlighted. A primary subject is the assertion of the developing countries to the effect that preservation of the global environment on a global scale is the responsibility of the developed countries, and that the developing countries should seek modernization by achieving economic growth giving low priority to preservation of the global environment, through exercising their rights to live and develop economically as fundamental human rights.   Under these circumstances where the world is changing greatly, the role of Japan in international society will also expand.

The role expected for Japan relies on the technological and economic power which Japan possesses.   Technology development will need to be continuously advanced, and economic power will also need to be maintained and developed. Assistance for the modernization of developing countries requiring technology transfer should be limited to assistance on the understanding that the global environment is preserved.   Also, as long as Japan renounces the use of military power, which was demanded fifty years ago by the USA, the present world superpower, Japan's strategy of cooperating with it and working together with the United Nations needs to be clarified.

Even by traditional macro-engineering standards, global environmental problems can be solved to some extent.   However, what macro-engineering aims at is the realization of large-scale public projects within   the constraints of global environmental problems.   For example, global transportation networks will be formulated that cover land, sea and sky (including outer space), harmonizing with the global environment and ecology systems and minimizing social friction.

THE SIBERIAN CORRIDOR BETWEEN EAST ASIA AND WEST EUROPE

The most important among global environmental issues for Japan at the present is the modernization of the Peoples Republic of China (PRC).   Development assistance must be given so that modernization will be carried out on the condition that the environment is maintained and preserved.   To make such assistance possible, it will be necessary to foster substantial use in the PRC of Japanese environmental technology and economic power.   It will also require Japan to deal with the former Soviet Union (FSU).

The present world economy has a still developing tripolar structure:

- the East Asia zone consisting of Japan, NIES countries and ASEAN,
- the North American Continent consisting of Canada, Mexico and the United States of America, and
- Western Europe including the EC and EFTA.

Assuming that the growth rate in real terms of East Asia, North America, and Western Europe would be 5%, 2.5%, and 3%, respectively (see Exhibit 12.3), the nominal GNPs in 1989 could be taken to be $3.5, $5.9, and $5.7 trillion, respectively.   On these assumptions, East Asia will catch up with North America and Western Europe in 2011 and 2015, respectively.

East Asia and North America, and also Western Europe and North America, are directly connected to each other through the Pacific Ocean and Atlantic Ocean, with high-speed container freighters supporting modern high-speed mass physical distribution.   However, there is no corridor to maintain such high-speed mass distribution between East Asia and Western Europe.   One hundred years before humans travelled to the moon, the Suez Canal was opened and Asia and Europe were connected, by which the world economy was greatly developed as population and production rapidly expanded.

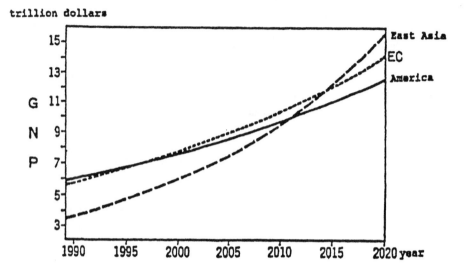

trillion dollars

**Ex. 12.3 Tripolar structure of world economy and
growth of nominal GDP**

What is sought now is a present-day equivalent to the Suez Canal project, that is, the development of a Siberian Corridor connecting East Asia and West Europe, in order to maintain and expand the world economy in a way that will contribute to solving global environmental problems. We need a railway transportation route to achieve high-speed freight transport. Such a project would require dynamic development along one of the following lines:

♦   a new surface railway with speed as high as the Japanese Shinkansen (bullet train);

♦   an underground tunnel railway, as proposed by Dr. Yutaka Mochida, who is now the chairperson of the Japan Macro-Engineers Society and who was a technical advisor of the Channel Tunnel, under whose leadership the Seikan undersea tunnel was completed which connected the main Japanese islands of Honshu and Hokkaido;

♦   an underground linear-motor car railway, as proposed by Dr. Yoshihiro Kyotani, who is now the chairperson of the research committee for large-scale transportation within the Japan Macro-Engineers Society and an originator of linear-motor car railway technology,

♦   a tube railway proposed by Dr. Frank Davidson, the founder of macro-engineering, through utilizing the present Trans-Siberian railway route.

Analogous to the company that constructed the Suez Canal, a business entity for construction and administration might be called, for example, the "All Nations (International) Trans-Siberia Railway Company." The company could be established as a multinational private organization, the capital being raised from respective nations and private sources. At present, there are reportedly ¥1,300 trillion (approximately $12 trillion) of financial assets owned by individuals and companies in Japan, only 0.2% or 0.3% of which would make possible the reality of a tunnel in the Siberian Corridor. Expansion of the two economic zones of East Asia and Europe would bring large profits to the company. The All Nations (International) Trans-Siberian Railway Company will surely be a key for the peaceful solution of global environmental problems through macro-engineering diplomacy.

The development of such a macro-transportation link encourages the movement of peoples, including even temporary movement such as sightseeing, business or studying abroad. It advances the distribution of information. Movement of peoples and distribution of information increase together; directly or indirectly they increase the number of people who are motivated by knowledge of modern culture, and they expand the large-scale migration of peoples.

HUMAN-MADE ISLANDS IN THE EAST MEDITERRANEAN

In Japanese the earth has been called by the beautiful name "a ballad of land and water". But throughout the world there have been many conflicts concerning land and water both within societies, and between societies, from the oldest times to the present. The fundamental cause was the increase of population due to the construction of the human-made environment. And the solution has been further construction of the human-made environment. Viewed on a global scale, the human-made environment, constructed recently through the development of modern science and technology, has made it possible to accommodate the huge population of the world. Every society has the basic responsibility to solve the problems occurring due to the increase of its own population within the global environment.

Seen from the Japanese perspective, the conflict between Arabs and Israelis, which has now become a world-wide problem, seems also to be originally a struggle to acquire land and water. If the population had been sparse, the problem would never have arisen. In this case, what is the solution of this international conflict caused by the construction of the human-made environment, if we are to treat the cause, not just the symptoms?

I suggest that macro-engineering diplomacy is now required. Land can be developed on the Eastern Mediterranean by constructing human-made islands and reclaiming land along the coast. Fresh water can be made from sea water. Recent developments in science and technology make it easier to achieve these. The scale of the enterprise would be decided dynamically through political decisions. The land area created on the sea would not be the same size as the land in conflict such as that of the West Bank and Gaza Strip, but the same productivity of land would be required.

The remaining problems are who proposes the policy to solve the conflict; which organization is the main contractor; and how to gather the money required. Both peoples involved in the conflict and other interested parties must be excluded. The main contractors should be chosen for their technical ability. The way to the solution is a political process aimed at treating the cause, as well as the symptoms. It is impossible to describe the way to the target in detail, because it depends on the situation. Dynamic macro-engineering is now required in order to decide the next steps.

CONCLUSION

In any event, frontiers between the developing and the developed countries will gradually be removed. Making common territory with developed countries will in fact benefit the economic growth of developing countries. Achievement of modernization for the developing countries will be easier.

In this brief chapter, I have offered suggestions for dealing with global environmental problems through the method of causal treatment extending over a long period of time. It is desirable that a prescription based on symptomatic treatment of the present global environmental problems be prepared.

In all events, we are looking to macro-engineering to offer reasonable solutions for saving the earth, and to macro-engineering diplomacy or dynamic macro-engineering to help solve the world's environmental problems, including regional conflicts.

13

Swiss Trans-Alpine Tunnel Projects

Herbert H. Einstein
Professor of Civil and Environmental Engineering, M.I.T.
Peter Testoni
Vice Director, Swiss Federal Office of Transport

A EUROPEAN PERSPECTIVE

Transalpine traffic plays an important role in Europe both for freight and passenger transport. In 1994, freight transported across the Alps by rail or truck amounted to 132.4 million tons.[1] This represents a tripling of freight transported 25 years earlier. Within the smaller arc from the Mont Cenis to the Brenner mountain passes, 84.9 million tons were transported in 1994. Exhibit 13.1 shows a reasonably balanced distribution between the participating countries but a severe imbalance regarding truck and rail traffic.

Particularly relevant is the fact that in 1970, 50% of all transalpine freight was transported through Switzerland, and in all countries rail transport dominated or was a least equal to truck transport. This mode shift between 1970 and today is similar to what happened in the U.S., and has a similar cause, namely, the opening of a number of limited access highways, including several transalpine road tunnels (Fréjus, Mont Blanc, Great St. Bernard, Gotthard, San Bernardino, Tauern). However, in Europe the reverse shift -- from truck back to rail -- has not been seen which took place in the U.S. in more recent years.

	FRANCE		SWITZERLAND		AUSTRIA		TOTALS	
	in 10^6t	in %	in 10^6t	in %	in 10^6t	in %	in 10^6t	in %
Rail	7.6	22	17.8	74	8.3	31	33.7	40
Road	26.6	78	6.2	26	18.4	69	57.2	60
TOTAL	34.2	100	24.0	100	26.7	100	84.9	100
% of total		40.3		28.3		31.4		100.0

Ex. 13.1 Distribution of Transalpine Freight Transport, 1994
(Mont Cenis to Brenner Passes)

The major reasons for the still relatively high proportion of rail traffic in Switzerland are the limits on truck freight transport (a 28-ton load limit compared to 40 tons in other European countries) and prohibition of truck traffic at night (from 10 p.m. to 4 a.m.) and on Sundays. These limitations to truck traffic were and remain an issue between the EU and Switzerland in its negotiations with the EU.

To compensate for these limits, Switzerland committed itself, in a December 1991 agreement,[2] to create so-called "piggy-back corridors" and construct two new transalpine rail tunnels to be opened in the decade 2001-2010 (see Exhibit 13.2). Creating the piggyback corridor along the same axes meant increasing the height of existing tunnels and then providing the required motive power. The latter is already in place in the form of new locomotives; the former was accomplished with the opening in 1994 of the Gotthard corridor for trucks (3.8 m high) and the opening in 1997 of the Lötschberg corridor for trucks (4.0 m high).[3]

Transalpine rail tunnels are also important for passenger traffic. Exhibit 13.3 shows the planned European High Speed Rail network. It does not include the most recent plans for Central and Eastern Europe; nevertheless, it is correct in showing several critical links crossing the Alps. The effect of such high-speed rail links on travel time is quite dramatic. For example, the new Gotthard rail line will reduce travel time from Milan to Zurich from 4.5 to 2.5 hours and the new Mt. Cenis rail tunnel will reduce travel time from Lyon to Milan from 6.5 hours to less than 4 hours. In other words, rail travel will become a competitive alternative to air travel. On the other hand, this means that the new tunnels will have to accomodate mixed traffic, which has consequences on capacity and safety.

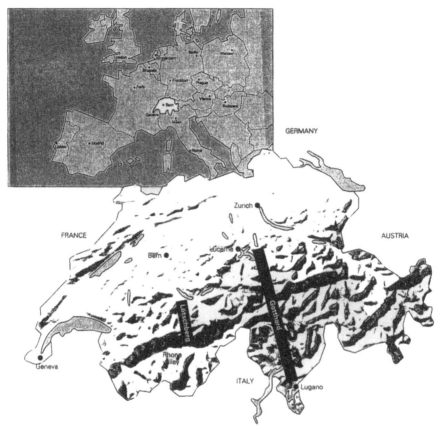

Ex. 13.2 **The Gotthard and Lötschberg corridors across the Alps. Both corridors are the location of the present rail lines, the piggyback corridors using these lines, and the new transalpine tunnels.**

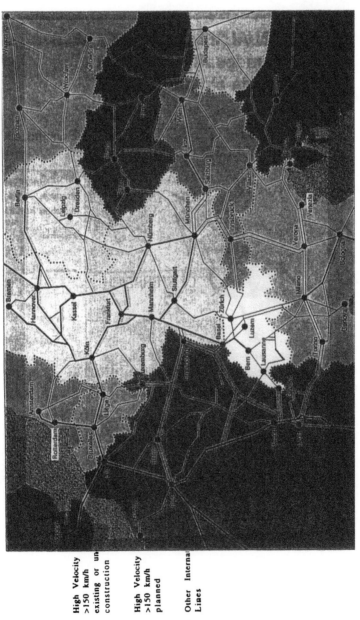

Ex. 13.3 Present and future European high-speed rail network

THE SWISS PERSPECTIVE

Historically, Switzerland has always played an important role in European north/south traffic and, to a lesser extent, in the east/west direction. While this role as a European "turntable" is beneficial from a purely economic point of view, it has undesirable environmental side effects. The creation of jobs, both from rail and road traffic, is outweighed (at least in the view of most people affected) by the negative inpacts of noise and air pollution. This led to the development and acceptance by popular vote of the 1994 "Alpine Initiative" which requires that the border-to-border freight transport crossing the Alps has to be by rail, and the capacity of roads in the Alps serving truck traffic must not be increased.[4] The shift from truck to rail has to be accomplished within ten years.

While furthering rail traffic is in the spirit of the 1991 EU-Switzerland Agreement, forcing a shift from border-to-border truck to rail transport is discriminatory and violates not only the spirit but also the letter of the Agreement. As such, these requirements cannot be directly implemented, both for Swiss constitutional reasons and because of relations with the EU.[5] However, by levying capacity-based fees on all truck traffic and special Alpine crossing fees on truck traffic independent of its origins, the same goal can be achieved without being discriminatory. As a matter of fact, the capacity-based fees correspond to the intention of the EU to create an equal cost basis for truck and rail transport, and they will make it possible to replace the 28-ton limit with a measure which better corresponds to free market principles. Most important is the availability of sufficient rail freight transportation capacity to substitute for truck transport. Hence the new transalpine rail tunnels will continue to play an essential role.

Shifting both passenger and freight from road to rail transport is not only important for the cross-Alpine traffic but for all ground transportation in Switzerland. This is reflected in the so-called *Rail 2000* concept which was formulated in 1985 and approved by popular vote in 1987. Its major purpose is to provide dense passenger service (i.e., one train per hour or every 30 minutes and less than one hour travel time between major nodes).

Implementation of the *Rail 2000* concept will require construction of new rail sections or improvement of existing ones in order to increase capacity and admissible velocity. The first stage of *Rail 2000* is presently under construction. However, its cost far exceeded what was originally planned. This fact, combined with a shift toward providing better connections to the European high-speed rail network, led to some changes in the concept. In essence, rather than building new infrastructure, improvements in railway stock such as double-deck trains and tilt trains (Pendolino) will be implemented. While originally separate from the financing of the transalpine tunnels, the continuation of *Rail 2000* is now part of the proposed overall financing model for public transportation in Switzerland.

THE HISTORICAL PERSPECTIVE

Transalpine traffic in Europe has existed since prehistoric times and was important for both the Celts and the Romans. It can be said that the founding of the Swiss Confederation in the 13th century was closely related to the transportation corridor over the Gotthard. As a matter of fact, only an engineering feat -- the construction of a suspended roadway along the steep walls of the Schöllenen gorge -- made the

Gotthard an acceptable route. Once this was achieved, it became one of the shortest European north/south routes. Nevertheless, this and the other routes across mountain passes were in essence mule paths.

The next major development occurred in the first half of the 19th century at which time roads were built over several mountain passes, making stage coach and wagon transportation possible. In the second half of the 19th century and the early 20th century, railroad connections were built across the Alps, first with no or only short tunnels and then, beginning with the Mont Cenis tunnel, including major tunnels (see Exhibit 13.4).

Tunnel	Year Opened	Length of Tunnel (in km)	Approx. Elevation of culmination (m.a.s.l.)
Semmering (Austria)	1854	1.5	900
Brenner (Austria)	1867	none	1350
Mt. Cenis (France, Italy)	1872	12	1350
Gotthard (Switzerland)	1882	15	1150
Simplon (Switzerland, Italy)	1906/21	20	700
Lötschberg (Switzerland)	1913	14	1250

Ex. 13.4 Major rail tunnels across the Alps

These rail lines (except for the Simplon tunnel) are characterized by culmination elevations of nearly 1000 meters or more. This means relatively steep grades as well as special alignment geometries, such as penetration into side valleys or spiral tunnels, to reach these altitudes. Both admissible velocity and capacity are unfavorably affected; all this is made worse by severe weather conditions in winter.

When the automobile and associated highway building boom fully affected Europe, it did not stop at the Alps, instead leading to the construction of a number of cross-Alpine highways, several of which included tunnels of similar length as the above-mentioned railroad tunnels, as indicated below.

Tunnel	Year Opened
Gt. St. Bernard	1964
Mont Blanc	1965
San Bernarino	1967
Tauern	1975
Gotthard	1980
Fréjus	1980

In the mid-1960s, Western Europe, including Switzerland, experienced rapid economic growth. One consequence was the highway building boom mentioned earlier. Also, rail traffic, particularly rail freight transport, expanded rapidly. Capacity limitations of the existing trans-Alpine rail links became apparent. A study group in Switzerland investigated a number of alternatives for a new base tunnel.

Such a tunnel would create a low-gradient alignment which would mean a substantial capacity increase for freight transport and a reduction of travel time for passengers. Various alternatives were studied and one known as the "Gotthard Base" alternative was selected. Field exploration and design were done from 1971 to 1973. However, the 1974 worldwide oil and economic crisis led to a major reduction in rail traffic as well as to tight financial conditions for the Swiss railroads and the Swiss government. As a result, work on the project was discontinued, although much of what was done at that time can now be used.

The 1980s and early 1990s brought similar economic growth and an even greater increase in traffic than occurred in the 1960s. The need for capacity increase, coupled with a desire to shift traffic to the environmentally more benign rail transport, led to the concept of *Rail 2000* and the trans-Alpine tunnel projects.

PLANNING AND FINANCIAL ASPECTS

In the mid-1980s transalpine transportation and particularly the construction of transalpine rail tunnels were reconsidered. This led to a number of planning documents out of which grew the so-called NEAT (*Neue Eisenbahn Alpentransversale* or New Transalpine Railroad Link).[6] The major components of the NEAT are the Gotthard and Lötschberg-Simplon axes but with additional improvements along the access lines (see Exhibit 13.5). The main axes follow the present corridors but with the two base tunnels, the 57 km long Gotthard base tunnel at an elevation of approximately 550 m a.s.l. rather than the present rail line's culmination at 1150 m a.s.l. and the 36 km long Lötschberg Base tunnel at an elevation of approximately 750 m a.s.l. rather than the present elevation of 1250 m a.s.l. The existing 20 km long Simplon tunnel at an elevation of approximately 700 m a.s.l. is a part of the new system. In addition, this includes several new tunnels of more than 10 km length along the access lines. The present capacity of 250 trains per day in each of the existing tunnels will be increased by 300 more trains which can use the base tunnels.

Clearly, this total tunnel capacity can only be utilized if the access lines are correspondingly improved to provide a four-track system in each of the corridors. While this was the case for the Gotthard axis of the original NEAT project, it was not so for the Lötschberg axis, mainly because the Lötschberg base tunnel had to serve as a replacement for an originally planned highway tunnel by providing a maximum of 132 trains per day for a road vehicle shuttle service. This meant that the combination of the existing tunnel plus base tunnel at the Lötschberg tunnel had to provide for an increased total capacity of 550 trains per day while a similar capacity increase on the access lines was not needed.

The total cost of this original NEAT concept was estimated at 14.9 billion Swiss francs (1991 price level). The financing plan was based on complete financing by the federal government. The necessary funds were to be made available in form of construction loans with interest at the level of the federal bonds. At completion of construction, these loans would be transformed into regular loans repayable over 60 years and at variable interest rates. The federal government in turn would take 75% of the funds from the general budget and 25% from the gasoline tax fund, the latter being justified by the role of the Lötschberg tunnel as a highway replacement.

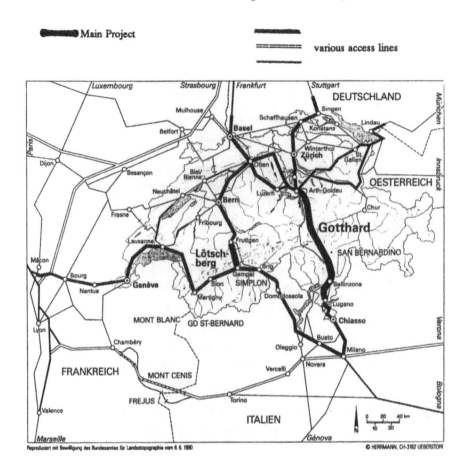

Ex. 13.5 NEAT project, including access lines

A review of this financing scheme by the federal government in 1994-95[7] and by the railroads that will operate the two tunnels (and thus be responsible for interest payments) provided some sobering information. the studies revealed that the originally planned NEAT would only be financially feasible if the freight transportation rates that the railroads can charge could be increased by 15% while simultaneously reducing the ratio of rail freight rates to truck freight rates by 10%!

The recent economic downturn in Europe and the oversupply in means of transport produced a complete decay of the freight pricing structure and made these rentability requirements unrealizable. In addition, construction cost increases seemed likely, largely due to increased demands for environmental protection which required additional tunnels along the access lines. Another reason for the cost increase is what usually occurs as projects enter the detailed design phase.

Also, other demands for enhanced public transportation -- in particular, rail transportation -- had to be satisfied. On June 26, 1996, the Swiss Federal Council submitted a proposal to Parliament to invest 30.3 billion Swiss francs (1995 price level) in railroad infrastructure projects.[8] Roughly 13.4 billion was intended for the trans-Alpine tunnels and access lines, 14.6 billion for other rail lines, and 2.3 billion for noise reduction along present and planned railroad lines. Financing would be provided through a combination of a gasoline surtax of 10 centimes and the previously mentioned capacity related truck transportation fees. In addition 25% of the NEAT portion of the investment is to be financed from the gasoline tax fund and 25% through debt in the form of interest-bearing, repayable loans to the operating railroads.

Concurrent with this change of the financing model is a significant reduction of the project cost of the NEAT by about 5 billion Swiss francs (1995 price level) mostly through a reduction of the access line improvements. Also, the Lötschberg base tunnel will be shortened compared to the original project, and only a 12 km long portion will be two-tracked (two parallel single track tunnels). A similar reduction of the project was also applied to the *Rail 2000* portion of the entire infrastructure concept.

This concept is, at present, in the form of a federal government proposal. It was debated, modified and approved by Parliament in 1996/97, and is to be voted on by the Swiss people in 1998. Since the original NEAT concept which was proposed in 1992 is still legally valid. Design and construction on those parts which will not change with the new concept are continuing.

BASE TUNNELS - TECHNICAL ISSUES

In addition to providing financial and political challenges, the new base tunnels are equally challenging from a technical point of view. The present design concept for both base tunnels consists of two single track tunnels connected by cross passages and rail crossovers (see Exhibit 13.6). The cross passages will, at least partially, serve as rescue features, and the rail crossovers will allow the operator to take tunnel sections out of service for maintenance purposes. The number and detailed design of the cross passages and rail crossovers is, at present, being studied. In contrast to the Channel Tunnel, there will be no parallel service/rescue tunnel, an issue which was resolved in the systems study.

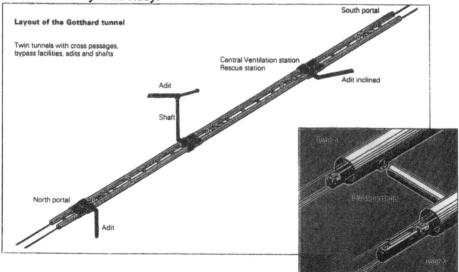

Ex. 13.6   Design concept for base tunnels (Two single track tunnels with cross passages, rail crossovers and ventilation/access facilities)

The tunnels will be under as much as 2000 meters of overburden (at the Gotthard perhaps even 2600 meters) which has a number of technical consequences. The rock temperature at such depths can be in the range of 50°C which requires air conditioning during construction and ventilation during operation. If the water table is at the terrain surface this means that water pressures of up to 20 MPa will be present in the tunnel. The rock overburden translates into vertical stresses in the rock of up to 50 MPa; similar or higher stresses can be expected in the horizontal direction caused by the Alpine tectonic movement in the north/south direction. These stresses are then increased by the creation of an opening. While good quality rock can sustain such stresses, this is not so for weaker material.

The great length of the tunnels combined with the slow advance rate caused by the difficult geologic conditions in some sections would lead to unacceptably long construction times if the tunnels were built from the two ends only. Intermediate shafts, or adits (refer to Exhibit 13.6) are required, although these will add to the cost. On the other hand, it will be possible to use these intermediate access points for ventilation and for rescue facilities during operation.

It is estimated that tunnel construction for the complete NEAT will produce between 30 and 35 million cubic meters of excavated material. There is literally no space in Switzerland to simply dump this, and an intricate scheme for reusing most of the excavated material is being developed. Reuse includes concrete aggregate, embankment material, or temporary storage and later reuse in quarries.

Significant issues regarding geological conditions in the proposed tunnel sites must also be considered. While the rocks in some locations are crystalline and of generally good quality with few faults, others are heavily disturbed and may be more soil-like. Encountering such conditions at depths of 1000 to 2000 meters below the surface is very problematic. Handling these conditions during tunnel construction will play a major role with significant effects on construction cost and time. In planning tunnels of such length, operational and safety issues are equally important. Determining construction cost and time as well as considering operational and safety issues are part of the ongoing detailed design work.

THE TUNNEL SYSTEMS FOR THE GOTTHARD TUNNEL

Both the Gotthard and Lötschberg tunnels are comprised of three types of tunnel systems (Systems A through C in Exhibit 13.7):

- a doubletrack tunnel with a service tunnel (System A),
- two singletrack tunnels with a service tunnel (System B), and
- three singletrack tunnels (System C).

Later, the fourth option with two singletrack tunnels (System D) was also included.

Ex. 13.7   Tunnel systems proposed for the new base tunnel lines at
Gotthard and Lötschberg
A: Double track plus service tunnel
B: Two single track plus service tunnel
C: Three single track tunnels
D: Two single track tunnels

Construction Cost and Time
As indicated earlier, geology governs the construction process, particularly if
geologic conditions are difficult. From a decision-making point of view, it is not
only the absolute cost and time but the associated uncertainties which are important
since they significantly affect the financial risk. A computerized procedure developed
at MIT, known as *Decision Aids for Tunnelling* (DAT),[9] allows one to simulate
construction of a tunnel at any desired level of detail and to represent the
uncertainties affecting the construction process. When using DAT geologists and
engineers determine all factors affecting construction and, in a formal questioning
procedure, estimate uncertainties.

DAT, which had been practically applied earlier, was modernized and transferred to Switzerland through funding by the Swiss Office of Transportation. The program was then used to estimate the construction cost and time of the Gotthard Base tunnel project. For this purpose, the tunnel was subdivided into so-called geologically homogeneous zones representing the geology in the area. In each zone characteristic geologic parameters such as rock type, fault zones, overburden and their uncertainty were estimated. A particular combination of geologic parameters will require that particular structural features be part of the tunnel construction, the so- called "support requirements". These requirements can then be directly translated into cost per unit length and subsequently into advance rates (tunnel length constructed per unit time).

The combined geologic and construction information, including the uncertainties, are used in the DAT simulation process, and the results of these simulations are scattergrams such as those shown in Exhibit 13.8. One sees the scatter clouds for each of the three tunnel systems, with three clouds for each system. These three scatterclouds per system reflect the influences of the various geological zones. As can be seen in the exhibit, the effect of the zones on cost is dramatic. The exhibit also shows that the doubletrack tunnel (System A) has the lowest cost but the longest construction time, while the System C, the three singletrack tunnel system, will run the highest cost.

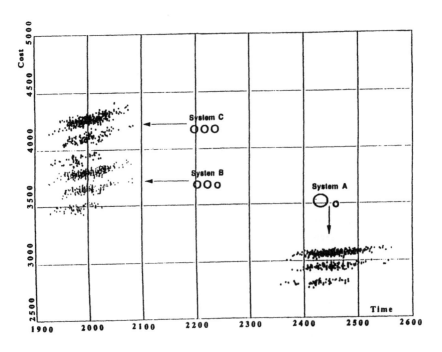

Ex. 13.8   **Gotthard Base Tunnel**
            Time-cost scattergram of the three systems
            (time in working days; 1 year = 300 working days)

Operational and Safety Aspects

Exhibit 13.9 shows the track arrangements for the three tunnel systems A through C. It is important to note that operationally, in a 50 km long railroad tunnel, there will be one maintenance/construction site requiring closing of the track each day all year long. Tunnel System C is thus operationally most desirable since one tunnel section can be taken out of operation while a two-track system still exists. Also, passing of running trains is possible and the number of switches is minimized, which is an advantage both from a maintenance and a safety point of view.

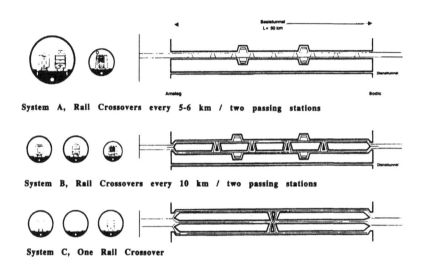

System A, Rail Crossovers every 5-6 km / two passing stations

System B, Rail Crossovers every 10 km / two passing stations

System C, One Rail Crossover

Ex. 13.9 Track systems for tunnel systems A, B, C

This system is also best *vis á vis* the safety of railroadworkers and passengers. The workers are isolated from traffic, and two rescue tunnels are available in case of a train accident. Ventilation of System C benefits from the train piston effect in single-track tunnels, but the additional artificial ventilation through fan-equipped ventilation stations is more substantial than in System B.

From a safety point of view, System A is highly unfavorable. Railroad workers have to work near a live track which leads to potential doubling of accidents compared to Systems B and C. Passengers in a disabled train are endangered by live traffic on the adjacent track, and fires cannot be isolated. Also, shifting loads on freight trains endanger oncoming trains. The operational advantage of a double-track tunnel is the relative ease of placing crossovers between tracks and thus having only short one-track sections during maintenance work. However, this goes hand in hand with reduced safety and increased maintenance associated with a large number of switches. Ventilation relies completely on artificial ventilation from ventilation stations.

System B takes an intermediate position regarding safety. It is operationally less efficient than System C, but it is best regarding ventilation.

System Selection

A multiattribute utility analysis considering the interaction of construction cost and time and of operational aspects was conducted. The utility associated with each system (shown in Exhibit 13.10) reflects the construction cost and time and their uncertainties on the one hand, and the operational and safety aspects on the other hand.

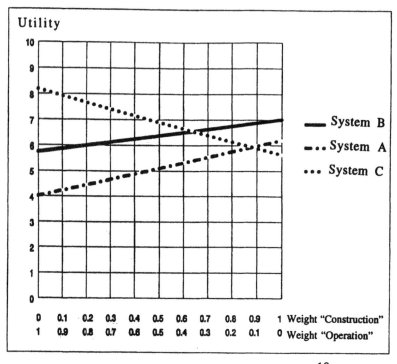

Ex. 13.10 Results of Multi-attribute utility analysis[10]

Considering actual construction costs and given the tightening financial situation, System C was considered to be too expensive. This led to the selection of System B and, based on additional studies, the elimination of the service/rescue tunnel, thus resulting in System D (refer back to Exhibit 13.7).

The service/ rescue tunnel was eliminated for three reasons. Operationally, the service tunnel contributes only minimally. The safety features provided by a rescue tunnel can be shifted to the cross passages and special rescue stations. Given these marginally small contributions, the cost of a service/rescue tunnel was not justified. In a small measure of justification, it seems that the two single-track tunnel system has now also been selected for the trans-Alpine tunnels in France, Italy and Austria.

A similar systems study was conducted for the Lötschberg tunnel, and it produced analogous results. The only difference between the two base tunnels is a somewhat larger cross-section of the Lötschberg tunnel (approx. 8.6 meters internal diameter rather than the approximately 8.4 meters at the Gotthard) to accommodate trains for the automobile/truck shuttle.

CONCLUSION AND OUTLOOK

Switzerland has committed itself to shift road traffic to the environmentally more benign rail transportation. This commitment includes the construction and operation of two new Alpine base tunnels. The lengths of 57 km and 36 km respectively, the overburden of up to 2000 meters, and the often difficult geologic conditions, make these tunnels extraordinary engineering challenges. Financing these transportation corridors together with other public transportation facilities also requires innovative financing and new resources.

Once completed early in the 21st century, these tunnels will become symbols of the new engineering which can handle extremely difficult technical problems while being environmentally and financially responsive.

NOTES

1. Eidg. Verkehrs-und Energiewirtschaftsdepartement (EVED), *Wege durch die Alpen*. Bern, 1996.

2. Schweizerischer Bundesrat, *Botschaft zum Transitabkommen zwischen der Europäischen Gemeinschaft und der Schweiz*, Bern, 1992.

3. Testoni, op. cit.

4. EVED, op. cit.

5. Testoni, op. cit.

6. Schweizerischer Bundesrat, *Botschaft über den Bau der schweizerischen Eisenbahn Alpentransversale*, Bern, 1990; Elektrowatt, Motor Columbus; Neue Eisenbahn-Alpentransversale, *Basisbericht* 24, Eidg. Verkehrs u. Energiewirtschaftsdepartement, 1988.

7. Coopers & Lybrand, *Financial Review of the Neue Alpen Transversale Project*, 1995. Testoni, op. cit.

8. Schweizerischer Bundesrat, *Botschaft über den Bau und Finanzierung der Infrastruktur desöffentlichen Verkehrs*, Bern, 1996; Testoni, P., *NEAT-Projekt der Schweiz Eine Standortbestimung*, Preprint, Interalp Verkehrsforum, 1996.

9. Einstein, H.H., et al., *Decision Aids in Tunneling*. Monograph, Swiss Federal Office of Transportation, 1992.

10. Schuster, P. & Kellenberger, J. *Doppel-oder Einspurtunnel am Gotthard*, Schweiz. Baublatt, January 1992.

14

Satellites in World Communications

William T. Brandon
Principal Engineer, The Mitre Corporation

INTRODUCTION

Underlying the revolutionary changes in communications is the ascendancy of information, which proliferates in form, format and dimensionality, as well as in quantity. As material goods have become more complex and unique, the associated detail endemic to their creation, manufacture, distribution, use and maintenance increases geometrically in response. Thus a particular screw used in an automobile engine may have a 12-digit part number to identify it uniquely from other parts.

Representation of information in digital form compatible with computers allows storage and manipulation of increasing volumes of information. Computers have been adopted in business and government, education, science and home use. Computer files of text, graphics, digital photographs, spreadsheets and data bases represent enormous quantities of information in compact form that is also convenient for electronic transmission. The computer has become an instrument of communications, an information appliance, with flexibility to handle many forms of information, including motion video.

We are entering an information age, characterized by information content provision, information delivery systems, and information appliances that allow man to connect.

INFORMATION APPLIANCES—THE HUMAN INTERFACE

Electronic communications may take place between humans, humans and machines (computers) or between computers. When the human is involved directly, an end instrument defines the form of the input and output. The telephone is the two-way voice instrument; facsimile machines transmit text; computers provide visual and audio and video capabilities. "Multimedia" refers to the combination of audio, text, graphics, motion video in an integrated presentation.

The end instrument, or information appliance, defines and enables varied forms and formats of information transfer, providing the human or man-machine interface. The existence of classes of instruments can be taken to imply their use, and hence the population of end instruments of various kinds provides insight into forms of communications and extent of usage. The personal computer is the first ubiquitous information appliance that can either function alone or as a communications interface.

Exhibit 14.1 presents quantitative estimates of the world's populations of communication instruments or information appliances -- telephones, cellular (mobile) telephones, facsimile machines, personal computers, and home broadcast television receive-only terminals. Data representing the period 1990-1992 and 1997 are listed together with an estimate for the year 2010.[1]

	1990-92	1997	2010
(per 1000 inhabitants)			
Telephones:			
America	6	29	45
Worl	10	12	23
TOTALS	(millions)		
Facsimile Machines			
America	6	15	40
Worl	14	35	120
Cellular Telephones			
America	6	40	80
Worl	11	89	300
Personal Computers			
America	70	160	250
Worl	100	220	400
Internet Users			
Worl	10	80	360
Home Satellite TV			
Worl	42	65	250

Ex. 14.1   Communications Indicators: Information appliances and special-purpose terminal population trends

Many other types of instruments could be included, and it must be noted that some new form of display or combination of formats may be introduced and become dominant. Particularly likely are three dimensional, naturalistic displays.

Telephones. The exchange of verbal information between individuals remains the fundamental human interaction, and the number of telephones per person has emerged as an index of communications sophistication, in turn believed correlated with productivity, gross national product, and average income. Since half the world's population has never made a telephone call, and the value of telephony is accepted, telephone systems are likely to be introduced first. In many cases mobile satellite systems will provide service rather than conventional wireline systems.

Cellular Telephones. Cellular phones are mobile or "wireless" phones that operate through a repeater usually located on a hilltop or building for wide coverage. Coverage is limited to line of sight, and a region is divided into "cells", each supported by a repeater. Repeaters also connect to the telephone system. Calls may originate or terminate at either a cellular phone or a desktop phone.

Facsimile Machines. The facsimile machine allows transfer of printed or text documents. The rise of facsimile parallels the rise of text in relation to voice as a preferred means of communications.[2] Facsimile transmission occurs via telephone lines which may obscure the true extent of its use. The combination of approximately 12 hours time difference, together with the value of examining, discussing and interpreting a text message in a foreign language, have so increased the use of facsimile between the West and Japan as to exceed the use of voice.

PERSONAL COMPUTERS AND MULTIMEDIA COMMUNICATIONS

The personal computer has rapidly proliferated since its introduction, fueling technology development that sees a two-fold improvement every 18 months. Due to its present position in terms of population relative to other information appliances, the personal computer is likely to be, for the foreseeable future, the most populous of the appliances now known. The projections in Exhibit 14.1 suggest a population of 400 million units by 2010, although this may be a conservative estimate. Even so, this quantity strongly suggests a market for multimedia communications. If there is sufficient growth of computers in areas of the world lacking ground telecommunications, satellite delivery will be successful. Even in areas where significant infrastructure exists, higher bandwidths and lower costs may enable satellites to capture a large share of the market.

COMMUNICATIONS INFRASTRUCTURE
Physical Transfer

Physical transport of information has always been the most prominent and perhaps pre-eminent method. Today, newspapers, mail, motion picture film, books, and computer software are among the types of information physically transported. But many forms of data may also be moved electronically. By 2010 many new forms of communications will be in use as alternatives to physical transport; indeed, physical transport may be all but eliminated.

An interesting hybrid is the physical movement of data stored in memory by a satellite in low earth orbit. The data may be received by a radio link from the transmit station and held in memory until the destination receiver is in view of the satellite, and then transmitted back to the ground. Termed "store and forward", this technique may be employed with a small, low-cost satellite. A single satellite can provide full earth coverage, but at the expense of time delays (up to hours, depending on locations of transmit and receive stations).[3] This may be overcome by using larger numbers of satellites. One system that is entering service in 1998 is

ORBCOMM which provides 2400 b/s data from hand-held terminals.[1]   The ORBCOMM satellite weighs only 98 pounds.   SATELLIFE uses satellites of this type in a medical network that provides consultative services to physicians in remote locations.

Telephony and Wireless Telephony

Conventional telephone systems involve substantial investment in the form of cable, right of way, switching centers and telephones, and they require considerable time to install. "Wireless" or cellular telephone systems are primarily developed for mobile users but may be applied to areas lacking conventional infrastructure. Both types of telephony (wired and cellular) will proliferate in underdeveloped areas. Digital cellular systems also support data communications to laptop computers.

Satellite communications may penetrate this market, especially where population density is low or geographic area is large. The ability of satellites to provide "instant infrastructure" (i.e., over a large geographic area) will be attractive and cost competitive not only for telephony, but also for many other applications.

Due to the cost of cable to the end instrument (sometimes called the "last mile"), a form of wireless known as "wireless local loop" may be employed and is another viable alternative to cable (wire or fiber), in addition to satellite communications. We expect the global application of satellite systems to have favorable economics.

Cable to Fiber Transition

Wire cable is used for both long-haul, high-capacity "trunk" lines and for individual lines to user phones. Intercontinental communications required undersea cable.

The first undersea cable enabled teletype (text) traffic between the U.S. and Europe. While this improvement in international communications is minuscule by today's standards, it should be recalled that from the early seventeenth century, transoceanic communications was limited to the physical transfer of documents by boat, and the infrequent crossings could require several weeks; thus, "news" was often several months old. It is somewhat surprising that telegraphy was the dominant form of communications by cable until the 1950s. The last telegram from France to the Cable Station at Orleans, Massachusetts, was transmitted in 1957, within months of the first artificial earth satellite orbited by Russia.

Transmission of analog voice required the use of electronic amplifiers spaced along the cable. This need led to development of long-life cathodes for vacuum tube electronic amplifiers in order to reduce the need for costly repair of undersea cables. The resulting improvements in traveling wave tube amplifiers were later critical to providing useful capability for communications satellites.

The technologies for transmitting information by pulses of light traveling through optical glass fiber advanced rapidly, and the first optical fiber cable was laid between the U.S. and Europe in 1988. Development improved the distance over which the light could be transmitted without regeneration of the light pulses and the speed of transmission or data rate was increased. About 1991, light of a single wavelength reached 2.5 Gb/s data rates in fiber. More recently, wavelength division

---

b/s stands for bits per second; kb/s for one thousand bits per seconed; Mb/s for one million bits per second; stands for Gigabits per second (one thousand million bits per second.

multiplexing, which uses multiple independent data streams of different wavelengths (or colors) multiply the communications capacity of a single optical fiber about ten-fold.

Fiber optic cables are being laid at astonishing rates since 1990. Major cables interconnect the continents. Fiber has completely supplanted wire cable in most long-haul applications. Countries lacking substantial telephone systems are in some cases going directly to fiber.

Fiber is fundamentally a *linear* medium, a cable that provides a point-to-point connection. At the terminus of a cable, a system is required for de-multiplexing the high data rate delivered by the fiber, and then delivering smaller data streams to geographically distributed users over cables or other fiber. Ultimately, wiring of some form is required to each information appliance. Installation of fiber on land by trenching has proven costly. While it is still actively pursued, the notion of fiber optic connection to any and all places on the globe may be many decades away, if it is ever accomplished. Fiber is excellent for long distance, high bandwidth trunks, such as links between continents that can be laid by ships without trenching (except on the continental shelf). These cables have tended to be placed between high population, economically well-developed centers, whereas the remaining world infrastructure problem is to bring electronic communications to all places and peoples. This is fundamentally an area coverage problem, for which satellite communications are well suited.

Satellite Communications

More than any other communications medium, satellite communication has dramatically and rapidly interconnected continents, nations, people, and cultures. The age of global communication by satellite was accurately foretold by Arthur C. Clarke in his 1945 article in *Wireless World* which described a concept of three geostationary orbit satellites delivering a global broadcast. The rate of change can be appreciated by considering that only five years elapsed from the first artificial earth satellite to the launch in 1962 of the first active repeater communication satellite; and the Communications Act of 1962 led to the formation of the International Telecommunications Satellite System (INTELSAT).

Given the state of global telecommunications in the 1960s, it was inevitable that satellite communications would first seek to provide global telephone services. Progress was rapid, aided by invention of the spin-stabilized satellite and advances in solid-state electronics (solar cells and transistors). INTELSAT I spanned the Atlantic in 1965; aditional satellites linked the continents by 1969, supporting thousands of simultaneous telephone calls between the United States and Europe. Other applications burgeoned: domestic systems (national rather than international, the first being for Canada); distribution of television; provision of mobile communications to ships, aircraft, and military systems.

Today we have entered yet another era in which personal communication of voice and digital data are to be offered on a global scale from satellites -- direct broadcasting of television to home or personal receivers; tele-education and tele-medecine; and support of computer based services, commerce and entertainment.

SATELLITE COMMUNICATIONS RENAISSANCE

Since the 1970s, satellite communications have been a dominant feature of global communications. The geostationary earth orbit (GEO) satellite has provided international, regional and national linkage where none had existed. Today there are 170 communications satellites in GEO. However, the rapid emergence of fiber optics has led to the notion that satellite communications might be in decline. During the 1990s, new satellite systems and services are in development that comprise a renaissance. The new systems will provide mobile telephone and data service between any two points of the globe. The next wave of innovation portends the provision of very high bandwidth fixed services to small terminals, for implementing multimedia communications services to business and home computers.

The U.S. National Science Foundation and National Aeronautics and Space Administration sponsored worldwide studies of satellite communications in 1992 and again in 1997, providing a unique assessment of the technology and global industry. Satellite communications continues to grow with market estimates of $30 billion in 1997 and as high as $75 billion per year by 2008 (market sales include satellites, launches, and terminals, but not resale of transponders, services, etc.) This continues to represent the biggest, best and perhaps only payoff from investments in space. A major globalization has occurred since 1992, marked not only by mergers and acquisitions but also partnerships and joint ventures. Proposed new systems have uniformly tended to be international in scope.

The World Trade Agreement has not yet fully opened communications markets. Some countries may yet insist on control over content but digital multi-media transmission renders this principle difficult to implement.

The major suppliers of satellites have become (or are becoming) service providers as a result of deregulation and the opportunity for new markets created by imaginative uses of technology. Organizations like INTELSAT now purchase satellites from competitors (who are now service providers on the international stage). Ironically, INTELSAT and the International Maritime Satellite Organization (INMARSAT), which brought international global satellite communications into being, face fierce competition from large consortia.

There is a shift in the manufacturing paradigm for satellites toward a more assembly-line approach. Several factories (IRIDIUM and Globalstar) have been implemented for specific programs designed for the shortest assembly time. Components and sub-assemblies are tested separately, then integrated without additional tests, in an effort to reduce laborious and costly comprehensive testing. These changes are concomitant with the increased scale sizes of systems (numbers of satellites) and global/regional definitions of markets, and desire to control costs and be "first to market".

New global players have asserted their presence (China, Korea, India and Israel) through major investments in facilities and announcement of products, further increasing competition. New satellite manufacturers in the U.S. include Motorola and Boeing. The net impact of all these changes will be a reduction in cost of manufacturing satellites, making satellite communications more cost effective.

New and emerging systems are directed at individual users -- in business, at home, or in travel -- thereby creating a mass market for satellite terminals. Mobile systems feature hand-held terminals; fixed service terminals are similar to direct (to home) broadcast television receive terminals. The mass market equates to millions of terminal units for both mobile/personal and fixed service systems. In comparison, the world population of very small aperture terminals (VSATs), typically of one meter diameter, is about 335,000.

Mobile Satellite Telephony

The first of the new wave of satellite communications systems will provide telephone connectivity between any two global locations with a portable telephone similar to a cellular phone. Hand-held terminals for mobile systems are just being unveiled. Although not yet in production, the systems anticipate the sale of a few million units. Initial models may operate only within the parent satellite system, but later models are expected to provide an additional mode compatible with at least one cellular standard.

The IRIDIUM system being developed by MOTOROLA will use 66 satellites in 780 km altitude circular polar orbits. Eleven satellites are spaced equally about each of six planes. Satellites are connected to each other by crosslinks with a plane and each satellite has two antennas to connect to satellites in adjacent planes. Calls are then routed from satellite to satellite or to one of a group of dispersed ground stations that connect to the local telephone system. Thus calls may originate or terminate in either an IRIDIUM handset, a cellular phone or a desktop phone. The IRIDIUM handset is designed to operate satisfactorily from inside a taxicab, a feature directed at the international business traveler.

Globalstar, a competitive system being developed by Globalstar Limited Partners, is based on using a different form of cellular telephone, called CDMA. Globalstar uses 48 satellites in 1414 km altitude orbits designed to cover areas below about 70° latitude. The satellites do not have crosslinks so communication is dependent on having a ground station within view of the satellite. Even with the higher altitude, this requires more ground stations than IRIDIUM; however, this feature brings in local ownership anticipated to smooth regulatory and other matters.

Several other systems are also in an advanced stage of planning and development. The various systems employ different orbits, numbers of satellites, and frequency plans, but all aim to support voice to mobile users throughout the globe using hand-held telephones. By 2010, this form of telephony should be commonplace. Many less-developed countries will use this form of communications instead of traditional land-based wired systems.

SATELLITE DIRECT BROADCAST

The particular form of satellite communications known as direct broadcast is potentially as significant as printing because it can distribute television throughout the world to hundreds of millions of people.

The first use of satellite communications for television transmission was for international distribution of programs and coverage of major events. It later became used for national distribution of programming by broadcast networks and then, in the United States, for distribution of programming to "cable head ends" (that feed local cable networks). Transmission employed analog techniques. Since broadcast stations and cable head ends could employ large antennas, there was no need for extraordinary satellite transmitter power.

Direct Television Broadcasting transmits the television radio signal from a satellite directly to a small home receiver. A large satellite could develop enough power to transmit several analog program channels. This type of service was pioneered in western Europe in such systems as ASTRA, based in Luxembourg. For British Sky Broadcasting, a satellite provided three channels doubling viewer choice.

Digital Television has been enabled principally by the development of techniques and equipment capable of compressing or reducing the number of bits required to represent the information in each frame of the color motion picture. In satellite terms, a transponder or single satellite channel may be able to support transmission of one or two analog television signals. The same transponder may carry ten or even twenty compressed, digital television channels, thus greatly improving the economics of satellite transmission.

Combining compression with high-power satellites allows the use of a small ground antenna, typically 18 inches in diameter, for reception. Since antenna cost is proportional to size, smaller antennas have resulted in affordable direct (satellite to home) broadcast, precisely as envisioned by Arthur Clarke. Future improvements will include higher compression and transmission of high definition television. The estimate of 250 million home receivers by 2010 (shown in Exhibit 14.1) is considered conservative and depends largely upon what develops in Asia.

Direct Audio Broadcast has been technically feasible for some years but its introduction has been delayed due to the need to harmonize its relationship to well developed traditional audio broadcasting. Direct audio broadcasting by satellite will provide many simultaneous programs with outstanding audio quality. A system will probably cover Africa before large scale introduction elsewhere.

Both audio and television direct broadcasting can revolutionize education. The extent to which these media proliferate will depend on the development of suitable programming or content provision.

WIDEBAND AND MULTIMEDIA SATELLITE COMMUNICATIONS

Even as the mobile satellite systems are being deployed, a second wave of new satellite systems intended to provide "bandwidth on demand" are being developed. These systems will employ higher frequencies known as Ka Band (30 Ghz uplink and 20 Ghz downlink). Both GEO and low earth orbit (LEO) are proposed and some systems combine both.

These systems can provide high bandwidth or data rates to small or modest sized terminals to support desktop video conferencing, internet service, multimedia entertainment and education, to homes or business locations. While more systems have been proposed than are likely to be funded, it is possible to describe a generic GEO satellite system. One satellite will support a population of small terminals, typically 0.7 meter antennas, receiving data at rates up to 50 or even 155 Mb/s. Typical access rates are likely to be near 10 Mb/s simulating an "ethernet in the sky". Transmission bypasses the traditional ground-based infrastructure.

Leading LEO systems include Teledesic (288 satellites at 1375 km) and Celestri (Motorola proposal involving 63 satellites at 1400 km altitude plus 9 at GEO). Lower altitudes are employed both to reduce the size (and therefore cost) of the earth station, and also to reduce the time required for signals to propagate to and from the satellite. For GEO (36,000 km), the round trip time is about one-quarter second.

LEO delays are smaller, proportional to altitude. The shorter delays become comparable to delays in long-haul fiber.

Delays are troublesome for high-speed transmission due to properties of current data transmission protocols. Changes in the protocols are being developed to make better use of GEO systems, but the LEO concept avoids the issue by reducing the delays.

Low-cost terminals will be the key to market success. There is consensus that for new fixed service, multi-media Ka Band systems, the home terminal cost should be less than $1,000 and the business terminal less than $10,000. To reach these costs, a large quantity is implied. This means that producers of terminals must produce in large lots initially to exploit the elasticity of cost with quantity.[4] These relationships are recognized and producers have proprietary schemes to realize the cost goals.

THE INTERNET

The Internet has grown at unprecedented rates of nearly 100% compounded annually when measured by the number of users having access or connection to it. With such wide acceptance, it might seem that its definition must be well understood and agreed, but in fact there have been controversies over definitions. An *intranet* is a local area network or a corporate-wide network. An *internet* is a network of networks. *The* Internet is the worldwide network that interconnects thousands of the world's intranets.

The Internet uses conventional transmission facilities. The interconnections are implemented in a hierarchical fashion through telephone cables, fiber optic cables, satellite systems, and other means. At the high end are network service providers (NSPs) that provide long-haul national or international backbone connections through cable or satellite transmission purchased or leased from carriers (e.g., AT&T, MCI, and Sprint). NSPs are interconnected to each other at multiple locations. Next in the hierarchy are internet service providers (ISPs), which serve as the bridge between NSPs and intranets or individual Internet subscribers. ISPs connect to NSPs using leased lines. Subscribers and intranets generally connect to ISPs using the telephone system.

As a connection is traced from an individual computer through the ISP and NSPs, the data rate of transmission increases. The possible number of routes or separate paths also increases. Typical telephone modems operate in the range of 2.4 kb/s to 28.8 kb/s, with 56 kb/s modems now coming into service. A company intranet will connect its computers through a private ethernet (10Mb/s) and connect (at one point) to an ISP typically at 1.54 Mb/s, termed the "T1 rate". ISP connection to NSP is typically a single T1 line, but may be through multiple T1 lines. NSP long distance lines may run at T3 (45 Mb/s). Fiber optic backbones may operate at 655 Mb/s, with data rates continually advancing with improved technology. In early 1998 INTELSAT announced a 45 Mb/s Internet satellite connection that will allow smaller countries a convenient means of implementating Internet service.

Thus a particular computer may be connected to any of millions of other computers worldwide through telephone, ISPs and NSPs. The connection is effected by a chain of devices known as routers.

Digital data files may be small (e.g., "come here, Watson", about 100 bits as text) or very large (e.g., large color photograph or digitized "x-ray", ranging in size from tens to hundreds of megabits). For transmission, data are divided into chunks called packets. Each packet is given the address of the intended receiver of the data. Each router looks individually at the address of each packet and makes an instantaneous routing decision, transmitting that packet to the next router along a path to the destination.

Because of the multiplicity of interconnections, any router may have more than one choice for routing a packet. Routers share information on connectivity and "traffic" (congestion), so that routing decisions may minimize transmission delay. Thus sequential packets comprising a particular file of data may actually traverse the network through many different paths and experience different time delays in transit. The packets are reassembled in the correct order at the receiver.

Since the "network" is actually comprised of many different routers, transmission media (cable, fiber and satellite), uses different transmission protocols, and may vary dramatically from packet to packet, one can appreciate the difficulty inherent in arriving at a succinct and satisfactory definition.

The World Wide Web (www.) is a way of accessing information via the Internet and may be considered an application or overlay. It should also be noted that the Internet serves more than the World Wide Web. Information is resident at computers connected to the Internet. Computers may have a home page or address that can be a starting point for locating related information.

The term "web" is appropriate because the home page may have a "hyperlink" (activated by clicking an icon using a mouse). Hyperlinks may be numerous and may connect to other computers in other locations anywhere on the Internet. Thus the hyperlinks form a web of interconnections. Of course, activating a hyperlink causes automatic communication between computers. The mouse click amounts to a request and the distant computer downloads or transmits information to the requestor.

Personal computers can retrieve information from or through the Web. The speed of the download is critical to performance. Information may take many forms (graphics, text, images, video, etc.), some of which comprise large files. If the data transmission rate is too small, considerable time is needed to download a file. By providing high data rates and short propagation time delays, Teledesic and Celestri have overcome these problems. A GEO satellite can support a large number of 400kb/s downloads to small (60 cm) antennas. The existence of many computers and the current limitation of telephone lines to 56 kb/s provide the basis for the anticipated market for multimedia satellite communications.

THE WORLD INFRASTRUCTURE IN 2010

The first generaton of multimedia satellite communications systems and second-generation mobile/personal satellite systems for telephony should be in operation before 2010. Fiber optic cables will crisscross major developed countries and connect continents and trade partners around the globe. Satellite communications will extend the Internet to all countries. Direct broadcast of television and audio programming will be available in most of the world. Communications in many forms will be available almost everywhere.

Satellite communications will have evolved to emphasize mobile and personal uses, direct broadcast, and public service applications (to education, medicine and safety). Telephony and video distribution may employ fiber optics in increasing relative amounts.

Satellite communications and fiber optics will reach a new harmony of complementarity rather than competition. Improvements in technology will include interoperable data communications protocol standards, used throughout the world (a major advance).

NOTES

1. International Telecommunication Union, *World Telecommunication Development Report 1996/97*, 1997.

2. Brandon, W.T., "Influences of External Digital Environment, Markets and Terminal Price Elasticity on Personal Satellite Systems", *International Journal of Satellite Communications*, November 1993, pp. 77-83.

3. Brandon, W.T., "A Data Courier Satellite System Concept", *International Journal of Satellite Communications*, December 1995 (re-publication of paper originally presented in 1973).

4. Brandon, W.T., "Market Elasticity of Satellite Communications Terminals", *International Journal of Space Communications*, Vol. 10, 1992, pp. 279-284.

15

Global Military and Civilian Telemedicine

John A. Evans[1]
MDPR Program Manager and Technical Director,
MILSATCOM, Hanscom Air Force Base

INTRODUCTION

The Medical Defense Performance Review (MDPR) was established in 1993 to help "reinvent" how health care is provided to U.S. military servicemen and their dependents. One of the MDPR initiatives has been rapidly to insert video conferencing, telemanagement, and telemedicine technologies to improve the quality and reduce the costs of delivering that care from major and minor medical treatment facilities, to wherever the need exists, for example, to patient homes and to remote military communities. The technologies and the processes now being reinvented have the potential to provide excellent access to quality health care anytime, anywhere.

This chapter concentrates on recent intra-regional telemanagement and telemedicine efforts and synthesizes key success factors essential for evolving self-sustaining global telemanagement and telemedicine networks for the twenty-first century.

PROJECT BACKGROUND

One prong of the MDPR effort was focused on developing a medical provider workstation. The prototype is being developed at Scott Air Force Base. The other prong was conceived as a joint civilian and military initiative focused on a user-evaluated and user-guided, phased deployment of computer communication networks emphasizing group and desktop voice, data, image, and video conferencing to support telemanagement and telemedicine. Concurrently emphasis was given to inserting high-value-added reengineered management and clinical processes based on collaborative experiences with best-of-breed leaders of civilian telemedicine and newly-empowered military users of telemedicine. These broad-based MDPR efforts recently received the Vice President Gore Hammer Award for excellence in reengineering.

The MDPR initiative was intended:

1) to improve the medical management, acquisition, operational and support processes within the Air Force medical community; and
2) to promote collaboration among the other military services, NATO, the United Nations, other coalition partners, and the civilian health care communities.

This top-down and outreach approach has successfully extended the enabling technology across regional medical organizations to evolve incrementally to a true telemedicine system serving health care providers and their patients.

The major constraints are technical interoperability and, especially, cultural issues.[2] Cultural issues most inhibit taking advantage of rapidly developing technology. To reap the full benefits of new technology requires reengineering of the organizational[3] and medical processes themselves. Health care providers and management executives who are experts in their fields are best positioned to redefine these processes and lead the changes.

VIDEO CONFERENCING

Video Conferencing (VTC) is essentially a television link between two or more locations. Most early successful systems were installed in government and large corporation conference facilities. The widening acceptance of VTC technology has lowered the cost of working and collaborating together on common projects and issues from remote sites.

VTC technology is now being extended to the desktop in personal computer systems. This has made the full range of capabilities found on personal computer systems (office automation, engineering, computing, file transfer and networking) inherent components of desktop video conferencing systems. Important among these capabilities is the ability to share computer applications, allowing conference participants to work on a shared electronic copy of documents such as budgets, reports, plans, engineering drawings, and multimedia patient records.

The market for VTC systems is experiencing explosive growth. This has been made possible by commercialization of technical advances in video and audio compression technology, by acceptance of internationally promulgated standards supporting proliferation of the technology, and by the increasing availability of low-cost dial-up digital communication services.

The high information content of television signals, typically computed at 90 Mbps, inhibited the development of VTC systems primarily due to the high cost of communication services. Contemporary compression technology enables effective interactive video and audio communication at data rates of 128 kbps, a compression ratio approaching 700:1.

Complete standards-based add-on systems for personal computers, including the television camera, an audio speaker, and communication network interface, are available today at costs ranging from $1,300 to $8,000. International standardization of video conferencing, increasing product integration and increasing production levels are expected to continue to cause prices to decline dramatically.

Worldwide dial-up digital communication services at a rate of 128 kbps are becoming increasingly available today through the ISDN BRI service offered by most telephone companies. The cost of ISDN BRI in most areas is comparable to that of two analog voice-grade telephone lines. The service is widely available today in western Europe and Japan and at many of the local exchange offices in the U.S. The service is expected to be a worldwide standard in the future.

PROTOTYPE TELEMANAGEMENT NETWORK

As a first step in meeting the goals of the MDPR, we focused on improving communication among top medical executives, the decision makers in the Air Force health care community, by acquiring a prototype telemanagement system. Most of the top executives are also physicians, thus providing early insight into the value of VTC, a key enabling technology supporting not only telemanagement but also later-to-be-deployed telemedicine. The prototype was installed at key locations to enable early user participation in the system design and to validate the system capability and performance prior to larger-scale deployment.

Early user participation also provided the opportunity for the real system experts, the system users, to work with and understand the possibilities of the technology, and then to reinvent how they do their jobs.[4] As we noted, only with changes in work processes can the full benefit of enabling technologies, such as VTC, be realized. Success of a prototype doctor-executive telemanagement system would also predispose support for continued infusion of the capability into the health care delivery community, thus metamorphosing to a true telemedicine system providing improved access and quality of care at reduced cost.

Data rates of 128 kbps are suitable for most medical applications. However, some future medical applications may require better video quality for video exchange. Consequently our system has been made scalable and suitable for incremental change to provide higher performance where and when needed.

The prototype system now extends to the managers' desktops, providing live dial-up VTC capabilities as well as application-sharing capabilities, allowing real-time collaboration on budgets, policies, and other management tasks. The prototype has been found by the users to reduce dramatically the time required to make decisions. In addition, decision quality has improved through increased access to subject-matter experts. Further, greater real-time staff participation in the decision-making process has created a better understanding of policies being promulgated. This has produced a greater sense of identification with new policies and hence better and quicker compliance.

The conference room and desktop VTC systems are installed at locations across the continental U.S., Hawaii, and Germany. Regional telemanagement extensions in progress include locations in Japan, Okinawa, Korea, Guam, Alaska and other countries in Europe, such as England, The Netherlands, Turkey and Italy.

In addition to the conference room and desktop capabilities, a twelve-site multipoint conferencing capability is installed in the Air Force Surgeon General's headquarters conference room in Washington, DC. The multipoint system allows twelve other locations around the world to participate in a conference, with each remote location being simultaneously seen and heard. This system enables the Air Force Surgeons General and their staff to interact frequently and in real time with the corporate executive group that runs the Air Force medical service, and to have them contribute to and participate in the decision-making process, resulting in better and more timely decisions. Regular meetings now include a Monday staff meeting conducted at 12:30 p.m. (EDT), with participants on-line in Germany at 6:30 p.m. local time and in Hawaii at 6:30 a.m. local time.

The primary communication services are provided by the ISDN worldwide network where available; otherwise, dial-up dual switched 56 kbps services are used. In addition, INMARSAT satellite services are used to reach remote areas, such as the United Nations Medical Field Hospital in Zagreb, Croatia.

TELEMEDICINE

As video conferencing is changing the way the world meets, telemedicine will change how health care providers are trained and how health care providers deliver medical services to their patients. VTC is a key enabling technology for telemedicine applications. Pioneering efforts to develop telemedicine capabilities have been done by Dr. DeBakey of the Baylor College of Medicine,[5] Dr. Jay Sanders, Director of the Telemedicine Center at the Medical College of Georgia,[6] and others.[7] Continuing medical education and consulting services are becoming increasingly available via telecommunications links. The future promises to include operations conducted on patients by remote physicians.

In support of the MDPR, we are now reaching out across various regional organizations toward health care providers and remote patients and adding telemedicine capabilities to existing systems. Our first application has been to fill in the specialty and subspecialty gaps at smaller hospitals and remote military communities. The concept for a prototype "Telemedicine Consultation Suite" allows a specialist at a remote facility to see and hear the patient, in conjunction with a general practitioner, nurse, or other health care aide with the patient. This was prepared by Dr. B. Hadley Reed.[8]

The initial MDPR clinical trials are driven by the needs of the users, and so are not "technology push" driven but are instead based on the business case analysis referenced above. The most promising specialty areas for telemedicine applications have been identified as allergy, cardiology, dermatology, mental health, neurology, pulmonary/respiratory, and ophthalmology. Under-utilized specialists in these areas are available at military medical centers.

The prototypes will be tailored to the needs of the provider facility and outfitted with suitable medical instrumentation to provide a remote specialist at the military medical center audio, visual, graphic (e.g., EKG) and other real-time medical data to enable an effective consultation with the patient and the local health care provider. The goals of the clinical trials are to assess the cost and benefits of telemedicine capabilities in the selected environments. Our measures of effectiveness are now being defined, but they will address the need to capture and document expected cost savings resulting from decreases in outpatient referrals and expected improvements in the quality of care. The results of the prototype evaluation should be able to guide planning for future use and deployment of telemedicine capabilities in the Army, Navy, Air Force and civilian communities alike.

We believe this medical technology has the potential for an even greater effect than the advent of the telephone, which enabled a local healthcare provider simply to call a remote expert with a quick question. The obvious limitation of telephone technology has always been the difficulty encountered by the local provider in adequately describing the situation to the expert over the phone without the clarifying assistance of pictures, graphics, or motion.

The ability to dial up the expert consultant in such a way that the consultant can see the patient with the physical finding, even hear the auscultation and view the x-rays, will dramatically enhance any provider's ability to deliver the best care to each patient fast with less cost for outside consults. Further, the local provider gains insights and expertise from the remote consultant, thus building greater teamwork between patient, provider, and consultant during the course of treatment, as well as specialized teaching for the provider whenever needed and/or relevant.

Other cost reductions include reduced patient lost work time. Reduction of lost work time includes travel to and from remote medical facilities, trips that today can take as long as five to seven days for a single appointment. In addition, the loss in work-related productive effort resulting from this travel is a more subtle, but significant, cost. Employers who are mindful of the full cost of the medical benefits they provide do not find these work- and travel-related costs trivial. Further, greater timeliness of treatment has always been known to have great medical benefit as well as to be a major cost reducer.

TELEMANAGEMENT AND TELEMEDICINE TESTBED
While group VTC technology is mature and to a large extent interoperable due to the international VTC standards, desktop technology is just emerging. Many different incompatible approaches are being pursued. Telemedicine technology is even more immature.

Proprietary video and audio compression techniques are being implemented by some vendors to optimize performance over local area networks, which are outside the scope of the current H.320 international VTC standard designed for ISDN services. Other vendors are using techniques to transmit out-of-band in a local area by using spare telephone wire, or by riding on the copper conductors that may be used for the local area network transmission media. Further, standards are not yet in place for applications sharing and other features now part of desktop video conferencing systems.

The desktop products also differ in terms of performance and features. Similarly, telemedicine systems which are based to a large extent on integration of other commercial off-the-shelf medical, video and audio components, also represent a new technology which is yet to be proven in use.

To address these kinds of interoperability issues and to evaluate and test candidate products, we have put in place a distributed interoperability and product evaluation testbed. This has:

- facilitated product evaluation and selection,

- provided a means to evaluate alternative systems' performance and stability,

- allowed early user participation in defining requirements, evaluating system concepts, and selecting products,

- helped vet potential system alternatives with selected users prior to deployment, which has facilitated providing the right technological tool to the user rather than mandating to the user the use of a technological tool,

- provided a tool for the project office to improve communication among the project team members, i.e., the customers and the project office, and

- functioned as an important tool for providing telemaintenance and teletraining support, thus reducing operations and maintenance costs.

PROTOTYPE NETWORK: THE CURRENT STATUS

Over the past four years a number of sponsors have augmented the efforts of the Air Force Office of the Surgeon General, our primary sponsor. Its funding was nearly doubled by others who saw the value of deploying a global network to support an integrated telemanagement and telemedicine network. This network leveraged the hundreds of millions of dollars being invested by commercial enterprises in the global information infrastructure. It also employed the talents of the Air Force's Electronic Systems Center and The MITRE Corporation which had over decades been developing and deploying command and control centers and interoperable networks that had similar characteristics to the telemanagement and telemedicine network needed by the Surgeon General.

The Surgeon General's telemanagement system has significantly enhanced the capability for making timely decisions. Annual savings have accrued not only in the form of improved decision making, but also in reduced management, training and medical travel costs, and reduced time away from primary duty stations, conservatively estimated at $3 to $6 million per year,[9] a half-year payback, which can be leveraged to extend the global network and further enhance the savings.

Awareness of the improvements generated by the inter-regional medical decision support capability has stimulated regional medical commanders to provide similar capabilities on an intra-regional basis. Supplementing the Surgeon General's regional VTC bridging capabilities, the intra-regional VTC bridging capabilities will dramatically reduce telephone costs and increase the flexibility of bringing doctors and medical executives together even while they are working apart. The enhanced decision support and cost savings generated by the system will intensify as the intra-regional bridges are brought on line.

Our challenge is to re-engineer the clinical decision support processes to facilitate communication between patients and their local doctors who can then collaborate with a distant medical consultant to create a close patient-doctor-consultant treatment team. This will result in quicker and more accurate treatment in cases requiring consultations, and will given experiences to the local doctor not unlike Continuing Medical Education (CME). An important testbed design feature stemming from this collaborative approach was to integrate the telemedical instruments into the physician's desktop computer along with video teleconferencing.

Our design contrasts with placing a telemedicine center hundreds of feet from the normal examining room environment. The technology-driven design of these large telemedicine centers has generally resulted in expensive centers, and limited use has resulted from the distance from the patient. These two factors, high cost and low usage, have reduced the actual and perceived cost-effectiveness potential of telemedicine. The cost-effectiveness of our efforts has been increased from this perception in a number of ways, two of which are: (1) to lower the equipment cost by scaling the equipment to the needs of the user as articulated by the user, and (2) to place the equipment where needed as articulated by the user, typically the examination room or the doctor's office.

Further, deployment of the testbed equipment has been such that existing consultative patterns are duplicated. This results in the least perturbation of already existing doctor-consultant relationships. Both fiscal and medical impact measurement techniques have been overlaid on the testbed trials in a fashion which will also allow comparison of results with other telemedicine trials.

In 1996 we also learned how to deliver telemedicine when the patients and the doctors are far apart from each other. The stage has been set to link isolated patients and care providers to distant, more capable care providers. This form of telemedicine will at first be used to provide better and more timely care for Air Force personnel in isolated locations or locations where limited military medical staff is available. As this technology and clinical practice evolve, we anticipate that better medical care will become available to many whose medical care is presently constrained because of distance between patient and provider. Beyond this, the interaction between the remote higher-qualified or more specialized doctor and the on-scene provider will enhance the on-scene provider's medical skills much more effectively than current approaches to CME.

NEAR FUTURE ON DEMAND, SPACE BASED DELIVERY OF MULTIMEDIA SERVICES

The thrust of these current efforts could help the Air Force, as well as NATO and transatlantic communities, to enhance current telemedicine initiatives by making them more mobile, to provide in-flight capabilities and to assist in consolidating, integrating and standardizing capabilities useful to the joint community.

Along with the Air Force initiatives, Army and Navy telemedicine initiatives are also rapidly emerging.[10] Ongoing telemedicine initiatives are available by accessing the Internet Home Page of the Office of the Secretary of Defense for Health Affairs *[http://www.ha.osd.mil/index.html]*. The major challenge will be concurrently to reduce medical and command and control fragmented development efforts while inserting new technology and to making the new capabilities interoperable and easier to use.[11]

Healthy cities and regional networks are paralleling military telemedicine efforts. Civilian efforts could leapfrog military efforts once the commercial communications infrastructure becomes more robust, secure, and less costly and once state licensing constraints are removed. Significant telemedicine efforts are currently under way at leading-edge cities and regions with extensive medical facilities, such as those in Massachusetts, Texas, Georgia, and Washington, DC. Each of these regions is also beginning to link with the other and reach out to global sites in significant need of health care.[12]

Annually, hundreds of billions of dollars are being spent by international consortiums to build and enhance a commercial global grid of telecommunications networks linking continents under the sea, on the ground and in space. Medical teams, as well as warriors and businessmen, should be ready to benefit from this huge commercial and military investment.

In space there is a new economic war[13] and satellite race[14] (see Exhibit 15.1) being driven by common needs for multimedia services of globally deployed military, businessmen, and travelers, rapid advances in infrastructure, and profitable communications products and services.[15]

Project	Function	Satellites	Cost ($B)	Lead Backer
Teledesic	High-speed data, teleconferencing	288	$9.0	Craig McCaw
Iridium	Voice, fax, paging	66	5.0	Motorola
SkyBridge	Data	64	3.9	Alcatel-Loral
Celestri	Data, broadcast, video	63	12.9	Motorola
Globalstar	Voice	48	2.5	Loral
ICO Global Communications	Voice	12	4.6	Hughes/Comsat

Source: WSJ, 6/17/97, p. A3

Ex. 15.1 The new satellite race (major players)

The replication, in various NATO countries, of easy-to-implement, low-cost, self-sustaining projects that leverage or link to the rapidly emerging global communications grid, is already familiarizing isolated NATO military units with basic store-and-forward clinical decision support services augmented as required by low-data-rate VTC to emergency rooms in NATO countries. This system, which is now in place, can be shared, adapted and enhanced to support the global needs of civilians and is indeed already doing so. Furthermore, as we attempt to expand NATO, and transition downward the American presence in Bosnia, we can put in place a "Medical Partners for Peace" service delivery system that, hopefully, will be used to reduce tension.

To achieve this vision and strategy, we acquired low-cost, commercial, off-the-shelf desktop/laptop computers with attachable medical devices and standards-based collaborative software capable of performing in low-cost store-and-forward Internet and ISDN VTC modes, with optional wireless and satellite capabilities for even more remote and aeromedical evacuation environments. Next, a phased worldwide telemanagement network was implemented. Subsequent steps implemented telemedicine and/or teletraining capabilities in Missouri, Nebraska, Ohio, California, Alaska, and NATO countries.

REVOLUTIONARY NEW SPACE BASED SERVICES IN 1998
This ongoing activity sets the stage for near future, on-demand, space-based delivery of multiple information services.

The basic concept of how these new services will be requested and delivered is illustrated by the "push/pull" concept. An information request or "pull" is executed by a remote warrior, medic, businessman, or tourist using a hand-held device similar to a cellular phone (e.g., Iridium subscriber set). The signal is picked up by a constellation of low earth orbiting satellites (Big LEOs) which use on-board processing and cross-linking capabilities to hop across satellites and down to the regional gateway nearest the desired destination. Here the signal is converted to a format understood by the telephone system and sent to the intended receiver.

If a response requires a multimedia high-speed broadcast, the return broadcast is sent to the appropriate mix of commercial and/or military direct broadcast satellites to the requester's earth station. This ground terminal could be a low cost, commercially available "lapsat" (e.g., direct PC) or a more complex and costly terminal, depending upon security and access issues or desired format of the request (such as text).

Motorola's Iridium capabilities will be globally operational starting in the fourth quarter of 1998. Shortly thereafter, US/NATO units, as well as businessmen, will be able to call in information "strikes" tailored to their warfighting, medical, or business needs anytime from anywhere.

The push/pull is handled through a space network ("Internet in the sky") similar to that just discussed. The return signal could be received by a phased array antenna on the skin of the aircraft and distributed via an on-board "server" to satisfy the differing needs of crew and passengers.

Space-based (direct broadcast satellite) delivery of television and data services to aircraft has already been demonstrated (e.g., in the Joint Warrior Inter-operability Demonstrations). On demand, space-based transmission of voice and data from aircraft (e.g., via INMARSAT and AFSATCOM) has been a reality since about 1980. Integration of the two functions into a push/pull service should not be a major undertaking. Such a system can shortly be delivering a variety of entertainment, business and aid services to airborne civilian and military travelers.[16]

THE FUTURE IS CONDITIONAL
In conclusion, this revolutionary delivery of robust space-based information services will soon be able to augment deploying and deployed contingency units. However, cost-effective, medically valuable services will depend upon how well we learn success and failure lessons from the already-deployed concept initiatives and the growing investment being made in high-cost, proprietary equipment acquisitions.

NOTES

1. The author is grateful to the following for their contributions: F.P. Davidson, J. Sanders, T.G. McInerney, W.T. Brandon, and L.M. Row.

2. Jowers, K. "Making Patients Feel Valued: DoD's Doctors Not There Yet, Experts Say," *Navy Times*, October 21, 1996.
    Excerpt: "Success is going to require a major cultural change in the military medical community, the [Military Health Care Advisory Committee] was told [during a briefing by the Department of Defense."

3. Evans, J. & Mekshes, M., "Reengineering of Medical Departments through the Defense Performance Review." *Proceedings of the National Forum: Military Telemedicine On-Line Today, Research, Practice and Opportunities*, March 27-29, 1995.

4. McInerney, Lt. Gen. (Ret.) T.G., first Director of the Defense Performance Review. A video interview, "On Reinventing Government," The Pentagon, April 1994.
    Champy, J. *Reengineering Management*. New York: Harper-Collins Publishers, Inc., 1995.

5. DeBakey,M.E., MD, "Telemedicine Has Now Come of Age", *Telemedicine Journal*, Vol. 1, No. 1, 1995.

6. Puskin, D.S. & Sanders, J.H., "Telemedicine Infrastructure Development", *Journal of Medical Systems*, Vol. 19, No. 2, 1995.

7. Preston, J., MD, FAPA. *The Telemedicine Handbook: Improving Health Care with Interactive Video*. Telemedical Interactive Consultative Services, Inc., 1993.

8. Reed, B.H., MD., "Prototype Medical Consulting Suite". Internal Memorandum, June 8, 1995.

9. Cost savings associated with telemanagement and CME networks, i.e., the Air Force Surgeon General network, the Army telemanagement network, Region 6 CME skills training network and emerging telemedicine networks.

10. Rich, S., "Battlefield Medicine Turns Electronic", *[www.Washingtonpost.com]*, October 18, 1996, p. A25.

11. McInerney, Lt.Gen. (Ret.) T.G., "The Command Tactical Information System: A Study of Evolutionary Macro-Engineering." In: Davidson, Frankel & Meador (eds.), *Macro-Engineering: MIT Brunel Lectures on Global Infrastructure*. Chichester, England: Horwood Publishing, Ltd., 1997.

Davidson, I., "An Introduction to Emerald City", a briefing published by Northrop Grumman; 1996.

"Global Command and Control System", a briefing published by Armed Forces Communications & Electronics Association, Fairfax, VA, 1996.

12. Telemedicine Research Center, "US Hospitals Extend Coverage to Caribbean," an article at [*http://www.matmo.army.mil/news/sections/civprog/carrib.html*], 1996.

Barron, J.H., *Atlantic Rim Network Newsletter*, Spring/Summer 1995.

13. Kim, J., "Motorola Plan Could Spark Space Wars," *USA Today*, June 18, 1997, p. 1B.

14. Hardy, Q., "Motorola Plans Another Satellite System," *The Wall Street Journal*, June 17, 1997, p. A3.

15. Gardner, J.N. & Barron, J.H., "Global Perspective: Telehealth at the Crossroads," *New Telecom Quarterly*, First Quarter 1997, p. 11. From a side bar in the article: "We estimate the market for new applications will grow to [a range of] $6 billion to $7 billion from $1.2 billion over the next five years."

Samols, M., *The information Imperative: Managing Care Means Managing Information*. Robertson, Stephens & Company.

16. "Phased Array Antennas Link Mobile Vehicles and Satellites," *Signal*, June 1997.

OTHER RELEVANT LITERATURE

Barron, J.H., Nielsen, H., & Sanders, H.E., "Telemedicine and the Atlantic Rim Network", a videotape, Atlantic Rim Network Productions, 1996.

Barron, J.H. & McWade, R.W., "Shaping Our Future: The International Boston Initiative," April 1993.

Brandon, W.T. & Evans, J.A., "A National Ka Band Satellite Communications System for the United States," *Third Ka Band Utilization Conference*, Sorrento, Italy, 15-18 September 1997.

Brecht, R.M., et al, "The University of Texas, Medical Branch - Texas Department of Criminal Justice Telemedicine Project: Findings from the First Year of Operation," 1996. See also, [*http://int-telemedicine.com/utartic.html*].

Evans, J., et al., "Case Study of Telemedicine and Distance Learning in the Army, Air Force, Navy and Department of Defense Health Affairs," *Telemedicine 2000 Symposium*, Panel Discussion, June 15-17, 1995.

Farmer, J.C., et al., "The Potential Uses of Telemedicine to Augment Critical Care In-the-Air," *Aerospace Medical Panel Symposium*, Rotterdam; September 29-October 2, 1997.

First International Congress on the Atlantic Rim, 11-13 November 1994. See also [*http://www.dac.neu.edu/ARN/ARN.html*]

Gore, A., "Creating a Government that Works Better and Costs Less," *The Gore Report on Reinventing Government*, 7 September 1993.

Heitman, R.K., "Multi-Lingual Translator of Service in Peace and War," *Hansconian*, 19 April 1996.

Hittle, A., "The Role of International Cooperative Research Development Agreements in Facilitating the Emergence of Telemedicine," *Global Telemedicine & Federal Technology Symposium*, 9 July 1996.

Hoffman, P., "The Provider Workstation: A Prototype that Helps the Provider While Collecting Patient-Level Cost Accounting Information," *Telemedicine 2000 Symposium*, June 15-17, 1995.

Hughes, S.E., "Medical Global Command and Control System (GCCS)," Aerospace Medical Panel Symposium, Rotterdam, September 29-October 2, 1997.

Kantor, R.M. *World Class: Thriving Locally in the World Economy*. Boston: Harvard Business School Publishing Corporation, 1996.

Maine Sunday Telegram, "Atlantic Rim Now Taking Shape," (editorial), 23 July 1995.

McInerney, T., "DOD Could Reap Billions in Savings Through Outsourcing", *Defense News*, 7-13 October 1996, p. 32.

Mekshes, M.M., Evans, J.A., & Hoffman, P., "Perspectives on Videoconferencing and Telemedicine: Application to Remote Communities", *Ocean Cities '95 Symposium*, 21 November 1995.

Nice Matin, "Télémédecine sans frontières," 22 November 1995.

Ocean Cities '95 Symposium, "Telemedicine Roundtable Discussion," 21 November 1995.

Providence Journal-Bulletin, "The Atlantic Rim Network," (editorial), 3 January 1996.

Transatlantic Telemedicine Summit. Brochure and Agenda, Boston, 20-22 May 1997.

Zajtchuk, R., "Research, Practice and Opportunities," *National Forum: Military Telemedicine On-Line Today*, 27-29 March 1995. Sponsored by the US Army Medical Research and Materiel Command in cooperation with the Association of the United States Army and Georgetown University.

Section Four

MANAGEMENT AND HUMAN RESOURCES

16

Starting a Macro-Project on the Right Foot

John H. Landis
Stone & Webster Engineering Corp., Boston

If there is an absolute rule regarding macro-projects, it is that they will surely fail if they are not set up to get input from all major stakeholders for project decision making, particularly early decision making, in a balanced and timely way. Many crucial decisions are made during the initial stages of such projects, and it is imperative that these decisions reflect the views and judgments of the main parties that will be affected as the projects progress. Faulty decisions early in the life of a project often lead to crippling troubles later on and irreconcilable differences among the major stakeholders.

Equally important at the outset of a macro-project is good basic organization: goals that are reasonable and well understood, yet stimulating and exciting; provision of adequate human, financial, and physical resources; efficient use of these resources; clearly defined lines of command; straightforward relations with and among major stakeholders; crystallized procedures tailored to the conditions, requirements, and environment (in the broadest sense) of the project; compre-hensive evaluation of risks and preparation of contingency plans; effective deployment of integrated quality assurance programs; and fulfillment of all government regulations.

A macro-project is like a hurricane in that minute differences in the guiding factors at any given point in time determine its future course. Way back in the 1730s, Alexander Pope created a better analogy: "Just as the twig is bent the tree's inclined."

The procedure to ensure that major stakeholders are adequately and appropriately engaged in helping to delineate a project's guiding factors hinges primarily on when and under what circumstances these stakeholders are brought on board, how they are treated, and the weight given to their advice. The initiators of a macro-project should develop this procedure carefully and completely, paying particular attention to the sequence of the steps involved and the instructions issued to all participants.

To help initiators cope with this and other challenges, the important steps necessary to launch a macro-project properly have been identified by the author, drawing heavily on the results of an informal National Academy of Engineering colloquy, and are presented here. The author stresses that these steps comprise only what should be called a "base case." This case will not apply *in toto* to specific projects, but all or parts of it may be used as starting points for development of protocols that meet individual project needs.

These steps have been arranged in seven categories in rough chronological order.

	CATEGORY
1.	Preparation of project plan
2.	Early implementation of project plan and development and delineation of "safeguard procedures"
3.	Financing of project
4.	Selection and indoctrination of prime contractor
5.	Setting prime contractor on defined course
6.	Resolution of differences among key stakeholders, including prime contractor
7.	Review and reconciliation of plans, procedures, and organization

In Category 1, the steps are quite simple and logical, but deviate from accepted current practice by involving major stakeholders meaningfully in development of a "master plan". First, the initiators (usually a small group of companies and/or individuals) should establish appropriate goals for the project. These goals should be spelled out in as much detail as is necessary to avoid confusion and uncertainty among the many parties engaged in or affected by the project, and they should be distributed to these parties as soon as they become known. Second, a conceptual plan should then be developed, using the criteria applied to the goals. Third, major stakeholders should be identified, classified, and contacted regarding their interest in the project. As the fourth step, those who have a legitimate interest should immediately be convened to review and modify the conceptual plan. This step may require several iterations and months of time, but its importance cannot be overemphasized and sufficient time should be allotted to do it right!

Based on their special knowledge and might, the initiators should then prepare the first draft of the master plan. This should be discussed promptly and thoroughly with all of the approved stakeholders and their comments taken into account. The plan should be redrafted and resubmitted to the stakeholders for concurrence and then published as a preliminary guide. Again, several months may be required for this fifth step.

In Category 2, the first step results from recognition of the fact that the master plan is a living document and must be nurtured and revised in conformity with actual experience throughout the term of the project. The approved stakeholders should be convened once more to develop procedures not only for implementation of the master plan but also for checking its validity as the project moves along and updating it to accommodate changes in the project. This is a difficult task, but vital to the future health of the project. More often than not, it is performed best under the leadership of one person -- someone chosen by the initiators with the concurrence of the stakeholders, but independent of both groups, whose knowledge of the project and objectivity are unquestioned.

At this point, the owner of the project should be formally identified. Usually there is a general understanding among the initiators and recognized major stakeholders of what the ownership structure should be from the outset of the project, but as the master plan unfolds the structure sometimes has to be altered. It is prudent, therefore, to delay announcement of this structure until the project has been well planned and reasonable stakeholder harmony assured.

The first act of the owner is to select his (or her or its) architect-engineer (A/E). The first act of the A/E is to study the re-drafted master plan and pinpoint the events, deficiencies, erroneous assumptions, mistakes, and other factors that might delay, complicate, or obstruct the project or might throw it off course. The A/E should then perform a preliminary risk analysis to evaluate the possible effects of these factors and develop preventive and compensating measures. These measures should include contingency plans to offset the worst of the effects if prevention does not work.

Next, the A/E move ahead with preparation of the general technical specifications and a preliminary cost estimate for the project. Upon completion of these tasks, the A/E should report the results to the owner and, with the owner's concurrence, seek preliminary regulatory approvals.

If these approvals are not obtained, the master plan and the general technical specifications should be modified to satisfy reasonable demands. If the demands are deemed to be unreasonable, the project should be stopped until a satisfactory compromise is reached.

If the approvals are obtained, the owner and major stakeholders should be apprised immediately of all changes that had to be made to achieve this objective. The owner might have to meet with the major stakeholders to resolve any complaints the latter may have regarding the changes.

In Category 3, the steps are again obvious and straightforward, albeit crucial. The owner should invite bids from competing consortia of investment bankers and, with the assistance of the A/E, evaluate these bids promptly. The owner should then check the winner's credentials with the cognizant government agencies and upon receipt of the appropriate "green lights" formally select the financial backer.

In Category 4, the owner takes the first step by selecting the prime contractor for the project and apprising the contractor in detail of the project's goals and master plan, much of which information the contractor will have previously acquired.

The prime contractor should next be informed in detail by the A/E of the general technical specifications and preliminary cost estimate that was prepared and review them. Upon completion of this review, the owner, A/E, and contractor should meet to resolve any problems, disputes, or differences in interpretation and clarify vague items.

Intertwined with these last three steps is an ongoing effort by the owner and his (or her or its) A/E to keep the prime contractor informed of all agreements made with regulatory bodies and major stakeholders.

The orientation of the prime contractor of course slips over into Category 5. The main thrust of Category 5, however, is that the owner and the A/E should work closely with the prime contractor to ensure that the project continues on its defined course. If the owner and the A/E give substantial advice and assistance to the prime contractor in the following four endeavors, the charted course will probably be followed throughout the project.

1. Developing detailed specifications for major systems and components.
2. Preparing requests for bids on major systems and components.
3. Evaluating bids and selecting key subcontractors.
4. Preparing new master plan and cost estimate.

There are only two steps in Category 6. The first step is to distribute the new master plan and new cost estimate to the key stakeholders for review and comment. The second is to answer the questions that will inevitably be asked and to resolve any significant conflicts that arise. Answering questions correctly and resolving conflicts amicably will require all the knowledge, sagacity, and diplomacy the owner-A/E-prime contractor team can muster.

Finally. the owner should sit down with his (or her or its) financial backers, A/E, and prime contractor to decide, on the basis of:

*   the new master plan,
*   the new cost estimate,
*   available financing,
*   regulatory agreements,
*   stakeholders' attitudes and current opinions,
*   quality of bids received,
*   adequacy of adopted procedures,
*   organizational structure,
*   perceived risks and benefits,
*   economic conditions, and
*   the political environment,

whether the project is still feasible and worthwhile. If the decision is "yes", the project should be pushed ahead as fast as possible. Time is money -- in the case of macro-projects, big money. If the decision is "no", termination efforts should begin at once, no matter how messy they may be.

Clinging to a project that is not properly planned and organized and not accepted wholeheartedly by its major stakeholders (particularly its potential beneficiaries), the cognizant regulatory authorities, and the political leaders of the nations involved, has been the downfall of many a macro-entrepreneur.

17

Leadership: Authoritarian or Collaborative?

Martin Barnes
Executive Director, Macro-Projects Association

A project is always a novel task, and it is usually to be managed by a varied team of people.   They are varied in their skills and cultures, varied in their levels of enthusiasm for the task, and varied in their levels of confidence in achieving successful completion of the task.   They need effective leadership.   If project management is management of change, ordinary management is management of no change, management of the *status quo*.   Self evidently, this type of management demands less leadership.

Many people still believe that leaders are born and cannot be made.   By inference, leadership cannot be either taught or learnt. I disagree. Leaders cannot be made from people who have no aptitude for leadership, just as musicians cannot be made from people who have no aptitude for music.   Equally, every proficient musician has been trained and has experience -- aptitude was not enough. It follows that we can develop leaders more quickly and effectively if we organize their training and experience.

There are two types of leadership - authoritarian and collaborative.

♦ **Authoritarian** leadership is characterized by such traits in the leader as "good in a crisis", "he makes quick decisions", "he always knows his own mind", "he is very decisive", and "he keeps himself to himself".

♦ **Collaborative** leadership is characterised by "nothing is a crisis", "decisions are thought through and discussed", "he is a good listener", "he is a good delegator", and "everybody's opinion is valued".

The history of engineering project management in Britain is populated by leaders of each type. Isambard Kingdom Brunel was a great engineering project manager of the first half of the nineteenth century who was an authoritarian leader and a very bad leader indeed. Most of his projects massively overran their cost and time targets and had awful technical problems when they were finished. He did not delegate badly -- he deliberately did not delegate at all. Consequently he had to make all design, construction, and commissioning decisions himself. There were so many to be made that he made all of them in a rush and none of them until they had become the largest current crisis. Near the end of his career, when he should have learnt better, he was project manager of the design, construction, and commissioning of the largest and most technically advanced ship of its time, the *Great Eastern*. At one stage the project was in a hopeless state of chaos, worry, and recrimination when, fortunately, Brunel fell ill. John Scott Russell, the shipbuilder, took over as project manager and soon had things going much better. Unfortunately Brunel recovered too soon and resumed command before the project was finished.

Working at the same time was another project manager, Robert Stephenson. He was a collaborative manager whose projects were technically and commercially as demanding as Brunel's, but Stephenson achieved almost universal success. His leadership method was characterised by selecting promising young assistants, developing them rapidly through closely observed accumulation of experience, and delegating decision making and leadership to them as soon as they could safely take it. Most of Stephenson's assistants went on to become great engineer/managers in their own right. None of Brunel's assistants did.

Two ancient examples do not prove a thesis, but perhaps the general models do.

THE AUTHORITARIAN LEADER

He is at the top of the project organisation chart and has delegated decision making to nobody.

As he makes all the decisions, he has a great many to make. If he is particularly good at organising his time, he gives more consideration to the big decisions than to the small but not enough to either.

He requires all the information about what is happening on the project to be sent all the way up to him. He does not have time to study it all so he studies and acts upon a small, randomly chosen sample of the information in the erroneous hope that this will convince everybody that he knows everything about what is going on and is influencing everything.

He mistrusts all his staff, as they appear to him always to be trying to undermine his authority. In fact they are only trying to gain a little influence over what is happening in order to sustain their own self-respect and the respect of those over whom they are set.

The authoritarian leader is never happy but he is particularly unhappy when he is not under extreme pressure. He has no way of making decisions except in a crisis.

His record of successfully completed projects is poor -- but then he has only worked on difficult projects.

The Collaborative Leader

He is at the top of the organisation chart but has delegated all decision making except strategic planning and selection of his own immediate staff to his own immediate staff.

He has few decisions to make and concentrates upon making them, having taken advice widely. He can take decisions in good time while there is still a range of options between which to choose.

He relies upon summarised information supplied by his immediate staff and assumes that they have their own information appropriate to their own levels of authority.

He makes a point of not intervening in decisions taken and to be implemented at lower levels. He is seen to trust all his staff by the authority which he gives them. He encourages them to seek his advice, as this allows him to influence their decisions without pulling back the authority which he has delegated to them.   The collaborative manager can be happy and be seen to be happy, as his happiness derives from his mastery of the project and his confidence in his staff. He can make decisions in a crisis but believes that every unforeseen problem is a fall from the ultimate in project management attainment. He and his team have a culture in which to be taken by surprise is a shortcoming which could have been avoided.

The collaborative project leader has a record of success -- but then he has only had easy projects to manage.

These two general models are described with exaggerated simplicity in order to emphasize the differences between them. Neither is often met in such purity in real life. Each project manager will form his own view about which model to adopt. I vote for the collaborative model. I have been using it in my own real life as a project manager and project management consultant for many years and my faith in it has never been seriously assaulted.

LEADING AND BEING LED

For every leader there are several led. It does not serve the interests of the project if those of us who are to be led make it difficult for those who are to lead. There are some simple rules for the led which will improve the quality of leadership around the project. It should not all depend upon the leader. Here are some of them.

♦ If the leader's decision is not clearly conveyed, ask and go on asking until it is. Play it back to him in your own words as a check on comprehension. One of the most "heroic" events in British military history was the charge of the Light Brigade at the battle of Balaclava in 1856. Six hundred cavalrymen charged to their death obeying a wrongly understood order. The culture in the British Army at that time was not to question an order, however stupid the order seemed.

♦ Give your leader your ideas on what we might do, and your data on what might happen if we do it, as part of his decision-making process. But do not be resentful if he chooses the option which is not your favourite. In particular, do not hold back in making his chosen plan work successfully.

♦ Seek your leader's advice when you are unsure about a decision which you have to take. If in doubt about whether you need his advice, seek it anyway. Nothing secures his confidence in you more than his belief that you will seek advice in all the situations where you might otherwise make a mistake.

♦ Report to the leader whenever you are significantly taken by surprise by something that happens. A new decision is necessary which he might well ask you to take, but if you do not report the new situation to him, his vision of what is happening will remain wrong. For the same reason, do not depart from an agreed plan without reporting it to him.

All these points of technique are applied common sense, but they are not always followed. Following them helps the team to work as a team and makes the leader's job easier.

ENGINEERED LEADERSHIP

By "engineered" I mean thought out, designed, and implemented. Nobody ever engineered authoritarian leadership. It is "intuitive", "born not made", "cannot be taught", "a product of two hundred years of breeding". Collaborative leadership can be engineered. It commends itself to trained engineers. It comes close to the "create by design" ethic which is the basis of engineers' training and of their professional activity.

On project management training courses, I have organised exercises in which young engineers experiment with different leadership styles and techniques and decide which are better than others. This helps to blow away the mystique surrounding the subject of leadership. Leadership is not a freemasonry, and leaders are not a race apart from ordinary people.

The engineered structure of leadership which seems to work best and to suit engineers well is the following.

We have a hierarchy of managers from the top level at which there is only one leader, the ultimate boss, down through many layers of lesser leaders to the lowest level where many people do their work with nobody under their command.

At all intermediate levels between the highest and the lowest, everybody is both leader and led. We are led by implementing the decisions communicated to us by our leader. We lead by making the decisions which our leader has delegated to us and communicating our decisions to those we lead. We pass information up to our leader to help him make his decisions, and we expect those we lead to pass information up to us to help us make our decisions.

Information systems should be designed to help with this structure of leadership and decision making. For example, do not generate large quantities of detailed data about what has just happened and shoot it all straight up to the top of the hierarchy. Do not generate comprehensive and detailed plans at the top of the hierarchy and shoot them down all the way to the bottom for implementation. If you do, you are assuming authoritarian leadership and will stimulate growth of all the problems which go with it. In particular you will leave the intermediate managers with no authority, and you will slide quickly into crisis mode.

Many project management information systems still have this unfortunate characteristic. The increasing power and convenience of computers over the last twenty years has made it too easy to generate and communicate more and more data. We have allowed volume of data and extent of transmission to dominate, and we have not followed value as the basis for information systems.

A simple device which I use in my project management consulting work is to ask people to describe their information systems. In something like nine cases out of ten, they describe the traffic in information, not the uses made of it. My next question is to ask people for examples of decisions which were improved by the use of information transmitted via the established systems. Also in something like nine cases out of ten, they cannot remember the last time they used the formally transmitted information in this way. I refer here particularly to information intended to facilitate control of time and cost. I conclude that much of the information which we laboriously assemble and transmit is unused.

My engineered leadership and decision-making model requires much less information to be transmitted and requires it to be transmitted over shorter organisational distances. The only information transmitted upward is that needed for decisions which will improve the plans for the remaining work in the project -- and it only has to go up one level. The only information transmitted downward is that of new decisions being communicated for implementation -- and it only has to go down one level.

Assume now that, around the hierarchy, there are many commercial and contractual interfaces between the different organisations contributing to the project. Every contractor, subcontractor, and supplier has his own leader and led, and all of them have differing commercial objectives. This dimension of realism makes the authoritarian model even less effective since the chain of authority is severely constricted at every interface.

The collaborative leadership model also has a problem when there are constrictions at the contractual interfaces. Project management leadership using the collaborative model requires that all the led will contribute data and possible solutions to problems to the leader, and that they will respond positively and energetically to new instructions emanating from his decisions. Traditional contracts discourage both these traffics across contractual interfaces. The contract provisions often make it in the contractor's commercial interest not to contribute positively to the buyer's decision making and to respond unhelpfully to new instructions. Adversarialism, not collaboration, is the norm.

This is why some of us working in engineering project management have sought to reform the contracts which are used. This has born fruit in the introduction of the New Engineering Contract family of contracts (the NEC). The NEC is the only standard contract which facilitates application of modern good practice in project management. When a buyer and seller sign up to the NEC, they have agreed to use the procedures for managing across the contractual interface which are set out in the NEC and which are a precise legal statement of the collaborative mode of leadership in action.

GRADUAL OR SUDDEN

The conventional view of how engineers become leaders sees the mass of engineers as engineers and not as managers, even less as leaders. The exceptional person is seen occasionally to emerge from the surface of the pond of engineers and to amaze his former workmates by his achievements in management and in leadership. We tell stories about his early career which show that you could see the signs that he was going to be exceptional, even when he was one of us. The conventional view is that only the few make the metamorphosis and that, when they do, it is sudden.

A different view of how engineers could and should become leaders is more helpful. As soon as a young person enters engineering activity of any type, they are led by the person who decides what they should do each day. They begin to experience being led on day one. The first time they are put in charge of another person, they begin to experience both being led and leading. If they have any interest in the applied science of management, they will begin to notice which styles and techniques of leadership seem to work better than others and modify their own leadership functions accordingly. Their development as leaders and the led is continuous from this point. It does not proceed at a uniform rate. It accelerates immediately after a promotion and slows down as the last promotion fades into the past. Over a relatively long time, however, the progression culminating in top leadership gradually takes place.

Those young people whose appetite and aptitude is for management and leadership veer toward those dominant skills. In just the same way, those young people whose appetite and aptitude is for design veer toward that as their dominant skill. To get to the top in any particular branch of engineering you have to be something of a manager and leader. To get to the top as a manager and leader, you need no other skill.

Adoption of this model of the development of leaders has two advantages. First, it stimulates the application of the collaborative style of leadership. Second, it identifies potential good leaders much earlier and accelerates their development. Engineering organisations that stick to the conventional model or to the authoritarian leadership style are just not cultivating the leadership resource within their own staff.

Although recent experience shows this, it is striking that it is also one of the main messages from the comparison of the leadership styles of Brunel and Stephenson, working one hundred and fifty years ago.

18

High-Performance Partnering Practices

Kathleen Lusk Brooke
George H. Litwin
The Purrington Foundation

INTRODUCTION: MACRO-PROJECTS ARE PARTNERED PROJECTS

There is no question about the utilization of partnering in macro-projects. Large-scale projects are, by definition, partnered enterprises because no one company can handle the design, engineering, construction, and operation of projects such as The Channel Tunnel. Can macro-engineering projects take advantage of the significant trend toward partnering in the field of engineering? Do the traditional approaches to engineering partnering apply, or does macro-scale demand a different kind of approach?

PARTNERING IS AN INDUSTRY TREND

Since the CII report *In Search of Partnering Excellence* (1991) and Cowan's (1991) study of the Bonneville Lock and Dam, Portland District, U.S. Army Corps of Engineers, partnering has begun to transform the practices of the engineering and construction industry. Partnering has been defined as: "A cooperative approach to contract management that reduces costs, litigation, and stress" (Cowan, 1991). As Weston and Gibson (1993) reported, projects that utilize partnering have less cost and schedule growth, fewer claims and change orders, and greater value engineering. Wilson, Songer and Diekmann (1995) cite such research in their discussion of partnering as more than a workshop, indicating that partnering is a change in the culture of the organization and, through the partnering process, the industry.

PROBLEM STATEMENT

There is a lot of activity between believing that the partnering process can work and driving it down through an organization composed of people with different motivations, backgrounds, individual agendas, and understandings of what partnering is. In macro-projects such implementation challenges are multiplied by the scope of differences in culture, beliefs, norms, and established ways of doing business. To prevent chaos, orderly and systematic ways of managing macro-projects will be required. New competencies must be learned. Old norms and practices will have to be examined and, often, discarded. New practices are needed.

To be successful, macro-project management has to be understood and bought into by all levels and by all the organizational elements, including suppliers, customers, financial sources, and top management. The success of macro-project management lies in its implementation.

In *Mobilizing The Organization: Bringing Strategy To Life* (1996), we describe the fundamental implementation processes that have led to successful cultural and structural change, and how they are utilized in large organizations, many of which operate through alliance and partnership. We have identified four systemic processes that are necessary for large-scale organizational change: Purpose, Infrastructure, Guidance, and Resourcing.

MOBILIZING PURPOSE

When the Queen Sirikit Convention Center was built in Thailand in preparation for the World Bank's 1992 meeting, there was a sense of shared purpose that created extraordinary performance. According to Khun Chuchawal Pringpuangkeo, CEO of Design 103 Ltd., there were no materials shortages, no time delays, no re-work problems, no claims and disputes. Everyone shared the same purpose, even though the project required the cooperation of many vendors, suppliers, and partners. The case in point had the national honor of Thailand as the underlying and unifying shared purpose (Lusk Brooke, 1993). But many projects do not have such an inspiring, cohesive goal.

The importance of shared purpose (Crowley & Karim, 1995) in partnering is well-documented. But perhaps the implementation of shared purpose can be strengthened. We have developed a method, very effective when merging companies or structuring the organization of large-scale, multi-partner projects, called a VTAC in which all the major partners meet off-site to agree on:

- Vision
- Key Strategic Thrusts
- Action Agenda, and
- Communication -- from Beliefs to Behaviors.

According to Uwe Kitzinger and Chris Benjamin of the Major Projects Association at Templeton College, Oxford (1997), a similar process is used with their engineering and construction firms convened. Up-front time dedicated to shared purpose strengthens design and lessens rework. Both André Bénard, Chairman of Eurotunnel (1997), and Graham Corbett, CFO of Eurotunnel (1994) comment that partner evaluation time spent at the beginning of a project is the most valuable and most overlooked source of performance improvement. VTAC is up-front time.

While many projects begin with a vision and action plan, it is in the communication of the vision and plan that success (or failure) can occur. Understanding must be established in two ways. There needs to be a clear translation from beliefs to behaviors -- a noticeable difference in the way partners now operate that is visible to the whole organization. Visible, measurable behavior means practices.

Second, there needs to be enrollment of a critical mass within the whole project organization -- a minimum of 25% of all participants: leadership, partners, managers, professionals, labor, administration, suppliers, vendors, government and other stakeholder partners, environmental stakeholders, end-users -- who are clearly committed to act in a different way to get a macro-project done on budget, on time, and in a way that really matters to all who take part. They must then enroll the rest of the people in their departments and project teams. Enrollment results in an emotional commitment of the organization, the evidence of the capability of "will" within the project (Genta and Sokol, 1996).

MOBILIZING INFRASTRUCTURE

With purpose agreed and a shared mental model of the project established (Saeed, 1994; Winch, 1993; Senge, 1990), a partnered organization requires a means of creating an organization and infrastructure that is temporary but which must function as efficiently as a more mature organization. Engineering projects have a jump start because the technical specifications recommend the required information flows. It can be helpful to hardwire the data flows, but it is essential to create an organization that enrolls and manages the people. We developed and utilize a process termed "chartering" that parallels movements within the engineering field.

Charters Already Included in the Partnering Process

At the conclusion of the traditional partnering workshop, the agreeing parties draw up their intentions in a charter. As sketched in the Associated General Contractors of America report (1991), the Partnering Charter presents Communication Objectives, Conflict Resolution System, and Performance Objectives. The charter is signed by the key project personnel including the general contractor, owner, subcontractors, and suppliers. The model charter underscores the commitment of "agreeing to agree."

Chartering -- With Teeth

Developed for PowerGen in the UK when the Central Electricity Generating Board privatized three power generation companies and spun out twelve electricity distribution companies, our approach to chartering cements shared purpose by defining goals, shared responsibilities, and individual accountabilities. First, the agreements must be shared, but then the plan must be translated into individual and team actions that go beyond the technical plan. What is the agreement for knowledge-sharing and project/organizational learning? What practices will the partnership establish for performance management? Will the information system be hardwired? How will feedback be tracked, communicated, and acted upon? The "teeth" in the charter developed by PowerGen produced significant performance results: in 1991 PowerGen and National Power went out of the starting gate as initial public offerings at the same price; by 1993, PowerGen was outselling National Power significantly.

Chartering acts as "intelligent quick-cement" for the project organization because chartering:

Maps the whole project system, including prime and subcontractors, vendors and suppliers, stakeholders, environmental groups, public and private partners, financing sources. Project mapping, utilizing dynamic analysis and modeling (Vennix, 1996; Saeed, 1994) adds clarity to complex delivery systems and puts in place the feedback areas and flows of needed cooperation;

Anticipates and pre-agrees the process to be used for change orders and claims, as well as environmental and other stoppages;

Establishes accountability for scope of work and fulfillment;

Hardwires groupware and other information flows, including organizational performance feedback;

Links management practices, selected for project excellence, to performance and reward;

Gains agreement to plans for resourcing, both material and human. Human resource planning should include training in required practices and feedback processes to provide performance and practices tracking to the management team and primary partners;

Chartering is purpose-made into infrastructure and action plan. This much is already being done well in partnering projects. But it is after this point that things can either get better or get worse, and where mobilization processes can bring about successful implementation.

MOBILIZING GUIDANCE
In a multi-organizational setting, everyone comes to the table with a different culture, different ways of doing things, different benefits to be reaped from certain courses of action. We might consider these gaps of differences like the synapses of a nerve system. Given this "hardwiring" of preference and value paths, isn't it to be expected that as time goes by, the synapses will widen and develop slack?

Practices measurement can quantify the "how" of organization, creating actionable feedback based on questionnaire ratings by subordinates and peers and providing the information needed to help managers strengthen their performance.

We have been engaged in practices research for a number of years (Litwin and Stringer 1967; Litwin, Bray, and Lusk Brooke, 1996; Lusk Brooke and Litwin, 1996) gathering organizational and practices data from executives, managers and subordinates in a wide variety of businesses. We have begun to see that managers are increasingly required to work with those over whom they have no direct control. In mergers, new teams suddenly have to work together. In project-based industries such as engineering, and construction multi-partner project managers are especially challenged by the need to influence many players, most of whom are allies or partners, not subordinates. Multi-partner projects are increasing in every industry, especially engineering. And in engineering, large-scale macro-engineering projects in which success may well depend upon organizational not technical factors, cooperative practices may represent a valuable future direction.

Therefore we will present data gathered in engineering and scientific/ technical industries where we have identified practices that make a difference in success. We will also present current expert data on cooperative partnering practices and identifying key trends. Concluding this section, we will present developing expert data on macro-engineering projects which may call for a specialized approach. In evaluating practices, we have asked experts to identify those that are "critical". As defined by Edwin Cameron (1994), the first Director of Project Management at Stone & Webster, "critical defines actions or behaviors which, if not taken or performed in a timely manner or if taken or performed incorrectly, would place the project and/or its lower level project elements in a crisis with little or no chance of a cost-effective recovery."[1]

In our discussion, we will present two kinds of practices: Management Practices described by direct reports, and Influence Practices used with partners, allies, and peers. In both cases, the practices that describe successful managers were generated by others -- direct reports, or described by top management, and by peers -- never by the managers themselves.

PRACTICES FOR PARTNERING SUCCESS

"All platitudes are not equal" might be the first observation to make concerning practices. All the practices presented, both those that validate and those that did not make a difference, are "good" and "positive." Bad managers don't last long enough to make it into our sample; so, we are surveying the difference between average and excellent performers. Our findings reveal that the difference between success and failure is too often not intention and not even positive action. Why, then, do some practices matter and others not?

We find that practices such as the ability to plan, while no doubt essential for engineering success, can be true for all engineers and therefore not a success differentiator. But the ability to communicate plans is a success factor. Therefore we might observe that in mobilization, it is how practices are seen and valued by others that matters.

As we take a look at such practices, we will explore whether or not different things matter to direct reports and other things matter more to peers and partners.

Management Practices -- Data From Direct Reports

Exhibit 18.1 presents a summary of management practices data collected from multi-disciplinary project management situations over a period of twenty years. Rated by direct reports, these data statistically discriminate between high-performing project managers and their average-performing counterparts. While the selection of practices came from subordinates, overall objective evaluations of project management performance were made by customers and top manage-ment who had no knowledge of our questionnaire data.

We also present in Exhibit 18.1 a group of management practices which did not differentiate between high-performing project managers and their average-performing counterparts.   All of the items in the exhibit might be considered "good management" and are typically taught in schools of management and management courses. However, these practices were not more characteristic of successful project managers -- both groups did them equally well.

Contrasting these two data sets -- Best Predictors of Success and Poor Predictors -- it is perhaps surprising to find that competencies such as:

- *Able to identify causal relationships and construct frameworks for problem-solving;*

- *Able to create realistic, comprehensive business plans;*

- *Sets specific deadlines for all project elements;*

are not those that distinguish success. Certainly, successful managers can be quite good at these things, but average-performing managers are just as good. It is not a quality that distinguishes the high performer. Instead, high-performing managers are clearly better at:

+ *Establishing clear, specific performance objectives WITH subordinates.*

BEST PREDICTORS OF SUCCESS	POOR PREDICTORS OF SUCCESS
Establishes clear, specific performance objectives with subordinates. **	Is able to create realistic, comprehensive business plans.
Helps subordinates understand how their jobs contribute to the overall success of the organization. *	Able to identify causal relationships and construct frameworks for problem solving.
Gives subordinates a clear-cut decision when they need one. **	Sets "specific" deadlines for all project elements.
Goes to bat for subordinates with higher management when the situation warrants it. **	Stands by management decisions, even if he/she disagrees with them.
Emphasizes and demonstrates goal commitment and persistence in achieving goals. **	Is loyal to the organization.
Gives subordinates feedback on how they are doing. **	Able to think years ahead.
Communicates his/her views to others regarding their performance honestly and directly. *	Is prolific in the production of new ideas.
Recognizes people more often than criticizing them. **	Holds regular staff meetings.

* $p < 0.05$ (confidence level = 0.95)
** $p < 0.01$ (confidence level = 0.99)

Ex. 18.1   Management practices that do and do not predict success in project management (practices rated by direct reports)

In fact, the practices that most helped managers be successful in terms of on-time, on-budget results had to do with relationship and feedback. Combining these elements with another success practice

+   *Emphasizing and demonstrating commitment to achieving goals*

calls up the essential definition of leadership, walking the talk -- in other words, being a living example of what is required and expecting others to live up to this standard. The management skills that count in complex situations are:

+   *relationship*
+   *feedback*
+   *involvement with your direct team*, and.
+   *leading by example.*

Influence Practices -- Data From Peers And Partners
Exhibit 18.2 presents a summary of influence practices data collected from multi-disciplinary project management situations over a period of twenty years. Rated by peers, these data statistically discriminate between high-performing project managers and their average-performing counterparts. While the selection of practices came from partners and peers, overall objective evaluations of project management performance were made by customers and top management who had no knowledge of our questionnaire data.

We also present in Exhibit 18.2 a group of influence practices which did not differentiate between high-performing project managers and their average-performing counterparts. Contrasting these two data sets -- Best Predictors of Success and Poor Predictors -- we find the most surprising result is that while:

- *In conflict situations, looking for points of reconciliation of views rather than differences;*

sounds like good partnering behavior, in fact the behavior that correlates with success is:

+ *Encouraging the open airing of problems and differences of opinion.*

Especially in engineering environments, early reconciliation of differences may subvert the testing of consequences essential for valid problem-solving. It is not good if everyone gets reconciled to a compromise solution and the bridge falls down. Another aspect to the airing of differences is that early reconciliation can damp down disputes that will later become delays, subversions, and work stoppages. Thorough airing and recognition of differences is critical to identifying and dealing with problems.

BEST PREDICTORS OF SUCCESS	POOR PREDICTORS OF SUCCESS
Knowing and being able to explain to others the mission of the organization and how it relates to their job. *	Being consistent with communicating priorities.
Understanding which decisions can be made alone and which decisions need to involve others. **	Making sure the role each person will play in accomplishing a task is clear.
Encouraging the open airing of problems and differences of opinion. **	Encouraging the reaching of decisions through a blending of ideas rather than through force.
Personally emphasizing and demonstrating goal commitment and persistence in achieving goals. **	Considering the views of others according to the logic of those views, rather than according to personal preferences.
Trying to influence others through knowledge and competence rather than through official status. *	Treating with an open mind requests to change plans and goals when circumstances seem to warrant change.
Responding in a non-defensive manner when others disagree with one's views. **	In conflict situations, looking for points of reconciliation of views rather than differences.
Behaving in a way that leads others to trust one. **	
	Insisting that group members make every effort to solve joint problems among themselves before taking them to higher management.
Being willing to share one's power in the interest of the overall organizational goal. *	Expecting others to find and correct their own errors rather than solving problems for them.

* p < 0.05 (confidence level = 0.95)
** p < 0.01 (confidence level = 0.99)

Ex. 18.2   Influence practices that do and do not predict success in project management (practices rated by peers)

Another significant finding in the partnering/peer practices is the emphasis on competence and knowledge in getting things done. We observe that:

+ *Trying to influence others through knowledge and competence rather than through official status* is more predictive of success than:

- *Considering the views of others according to the logic of those views, rather than according to personal preferences.*

The distinction is subtle; both practices advocate non-personal criteria for judgment of ideas. But in engineering partnering environments, it is not necessarily logic that prevails but competence and knowledge. Perhaps this point relates to the earlier finding just discussed above concerning airing of differences as a way of testing ideas. Logic (poor predictor) may not be the definitive approach if the argument is rounded on a shaky premise; testing will disprove or validate the premise.

Critical Practices For Macro-Success -- Expert Findings
We further validated our findings by presenting them to a group of engineering, scientific, and academic leaders who reviewed the practices that are predictors of success. Exhibit 18.3 describes the expert views on which practices, compared with practices rated "not critical," are essential to achieving excellence in macro project management.

Of the eight practices rated critical to excellence in macro-project management, five are from those which have statistically differentiated between high-performing and average-performing project managers (refer back to Exhibits 18.1 and 18.2). The other three practices can be considered supportive; high-performing project managers tend to score higher on these, but the differences are not statistically significant. The supportive practices selected by experts on macro-project management as "critical" include:

+ *Recognizing subordinates for innovation and calculated risk taking*
+ *Providing help, training, and guidance for subordinates*
+ *Willing to make tough decisions in implementing the overall strategy*

In contrasting those practices rated critical with those considered not critical, the themes seem to be goal/decision focus versus interpersonal relationships/process. The critical practices are goal- and achievement-related, with some consideration to providing help and building trust. The noncritical practices are influence practices having to do with cooperation, friendly relationships, non-defensiveness, and participative decision making. Clearly, the experts see goal focus and strong decision-making capacity as the attributes necessary for macro-project leadership.

PRACTICES RATED CRITICAL TO EXCELLENCE	PRACTICES RATED "NOT CRITICAL"
Clear Decision Giving subordinates a clear-cut decision when they need one. (Management)*	Helping subordinates understand how well jobs contribute to the overall success of the organization. (Management)
Objectives Establishing clear, specific performance objectives with subordinates. (Management) *	Emphasizing cooperation versus competition. (Influence)
Commitment Emphasizing and demonstrating commitment to achieving goals. (Management)*	Noticing and showing appreciation for extra effort. (Influence)
Recognition Recognizing subordinates for innovation and calculated risk taking. (Management)	Building warm, friendly relationships. (Influence)
Help Providing help, training, and guidance for your subordinates. (Management)	Responding non-defensively when others disagree with your views. (Influence)
Mission Knowing and being able to explain to others the mission of the organization and how it relates to their jobs. (Influence) **	Making sure that there is a frank and open exchange at group meetings. (Influence)
Decisions Willing to make tough decisions in implementing the overall strategy. (Influence)	Deciding when decisions can be made alone and when you need to involve others. (Influence)
Trust Behaving in a way that leads others to trust you. (Influence)**	Evaluating views based on knowledge and competence versus organization position. (Influence)

*　differentiated between high-performing and average-performing project managers as seen by direct reports (see Exhibit 18.1)

**　differentiated between high-performing and average-performing project managers as seen by peers (see Exhibit 18.2)

Note: Headings under "Critical to Excellence" are not part of the practice and are shown only for the purpose of reference to Exhibit 18.4

Ex. 18.3 Expert selection of practices critical to excellence in macro-project management versus practices considered "not critical"

As Exhibit 18.4 shows, there is a consistent difference between average-performing project managers and their high-performing counterparts on all of the eight practices selected by macro experts. The differences are smallest in the middle of the graph, in the areas of recognition and help -- where both groups are near to average as compared with a multi-national norm base.

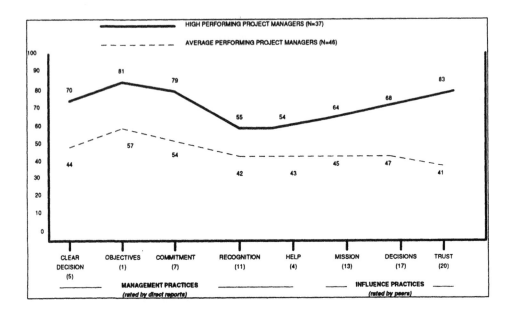

Ex. 18.4   Validation of expert selection: How practices relate to performance success in multi-disciplinary project situations

The biggest differences occur on the left and right sides of the graph. On the left side, with regard to management/direct report practices, significant differences are shown for *clear decisions when needed, setting clear objectives with subordinates*, and *demonstrating commitment and persistence in achieving goals*. On the right side, with regard to influence/peer practices, the most significant differences occur in *behaving in a way that leads others to trust you*, followed by *ability to make tough decisions when needed*, and *knowing and being able to explain the mission*.

Macro-Projects: Are They Different?

By definition, projects such as the Channel Tunnel are multi-partner. Do the practices required for partnering apply to macro-projects?

We asked a team of engineering partnering experts to select those practices that are critical for success in multi-disciplinary, multi-national partnering projects. In Exhibit 18.5 we contrast the top four practices selected by experts on partnering and by experts on macro-projects.

PARTNERING TOP 4 PRACTICES	MACRO-PROJECT MANAGEMENT TOP 4 PRACTICES
Knowing and being able to explain the mission and how it relates to people's jobs. (Influence)	Knowing and being able to explain the mission and how it relates to people's jobs. (Influence).
Willing to make tough decisions in implementing the overall strategy. (Influence)	Willing to make tough decisions in implementing the overall strategy. (Influence)
Behaving in a way that leads others to trust you. (Influence)	Giving subordinates a clear-cut decision when they need one. (Management)
Giving subordinates feedback on how they are doing. (Management)	Emphasizing the demonstrating commitment and persistence in achieving goals. (Management)

* Experts on macro-engineering who selected these practices as critical predictors of success include:

André Bénard, Eurotunnel; Ranko Bon, Department of Construction Management and Engineering, University of Reading, England; Edwin Cameron, Functional Management, Carver, MA; Janet Caristo-Verrill of Macro-Projects International, Wayland, MA; Y.H. Cheung, Apex Project Management Company, Hong Kong; Joseph Coates, Coates & Jarratt, Washington. D.C.; Leonard J. Delunas, Intercon Group, Bangkok, Thailand; Frank P. Davidson, Macro-Engineering Research Group, MIT, Cambridge, MA; Peter E. Glaser, Arthur D. Little, Cambridge, MA; Cordell W. Hull, formerly of Bechtel and present Chairman of Energy Asset Management, San Francisco, CA; Wallace O. Sellers, Enhance Financial, New York.

Ex. 18.5 Top 4 practices ranked by experts on partnering and macro-project management

Partnering Requires Feedback. Partnering experts, not surprisingly, selected 75% Influence practices. But even the partnering experts' one Management practice, on feedback, was transformed by their emphatic insistence that feedback be:

+ *Two-Way, with subordinates giving feedback to management;*
+ *Timely, with information following closely on results;*
+ *Faster, with less space between data and communication.*

Performance goals were seen to best be cooperatively established. There was great emphasis on joint goal setting and participation. Additional comments linked the performance review process to continuous improvement. Partnering experts recommended "devising methods of compensating people using measures which integrate personal, organizational,and multi-organizational outcomes."

Organizational learning, risk taking, innovation, and the encouragement of these by managers, was also seen as important by partnering experts. A combination of both performance and feedback, "recognizing subordinates for innovation and calculated risk taking" was seen by some experts as "coaching of the intellect." One expert noted that practices need to reflect the navigation of the unknown that is dictated by change.

Macro-Partnering Emphasizes Clarity and Leadership over Feedback. While macro-projects are inevitably multi-partner projects, the partnering can be so complex that experts stressed the need for clarity and leadership. For example, both partnering and macro-experts selected the same practice:

+   *Knowing and being able to explain to others the mission of the organization and how it relates to their jobs*

but for macro-experts it was their first-ranking priority, with a 70% consensus. Macro-engineers made no comments on the distinction between "mission" and "vision" (as did experts on partnering projects of smaller scale) -- perhaps because macro-engineering projects such as the Eurotunnel already convey quite a lot of vision. However, linking up the project with the organization and the individual job was seen the most important action managers can take in macro-environments. In fact, Dr. Yan-hong Cheung, of APEX Project Management Company Limited of Hong Kong and Shanghai, selected only this practice as critical.

Another significant contrast is the role of feedback. While partnering experts selected feedback as their number one priority, with a 95% consensus, macro-experts valued feedback much less (30%).

Managing down the chaos in macro-engineering was a priority macro-experts identified as critical:

+   *Giving subordinates a clear-cut decision when they need one.*

Again, no comments concerning a more participative approach from macro-experts, indicating, perhaps, that in projects as complex and potentially confusing as macro-scale initiatives, management excellence is achieved by being very clear. Such strong, directive leadership was cited by Frank Davidson (1983), author of *MACRO* and original proposer of the Channel Tunnel, as the core quality of successful leaders and mobilizers like Tallyrand.

Leadership and clarity -- the two touchstones for the macro-experts -- underscore Davidson's observations, making the kind of leadership described by James MacGregor Burns (1978) relevant for large-scale endeavors.

The Future of Practices Feedback In Engineering

Our results indicate that management in engineering is changing, with participative approaches on the increase. Influence practices were ranked as highly important by smaller-scale and large-scale experts, with influence being 75% of success in the former and 50% in the latter. But new management practices -- interactive practices -- emerged in the partnering rankings, especially clustered around feedback and trust.

It is likely that the increase of partnered projects in engineering will continue to emphasize participative feedback and trust.   Those projects and companies that emphasize such practices are more likely to be successful in partnered projects.

In the expanding realm of very large-scale engineering projects, while clarity and leadership will continue to be essential, we might expect that more participative methods, such as chartering, will begin to appear. We might anticipate that participative partnering processes will become important in macro- projects at the delivery levels, coordinating actions between local partners. Critical practices can be established at the beginning of projects, guiding performance management systems that link the many partners.

What to do with Practices Once Selected
Just as in the military where guidance systems are the feedback-driven means of conveying information, commanding and controlling operations, so guidance must be the cohesive "skin" of the mobilized organization.   To create consistent guidance, we utilize practices as the customized heart of a feedback system that provides ongoing data to managers and teams in two stages:

A. *Establish Required Practices*
When the partnered project is launched and chartered, we conduct a survey to see what the project requires in terms of behavioral criteria.   Practices seen as critical -- either because they are universally agreed as essential for success or because they are lacking and need to be added and reinforced -- then become the framework for the practices feedback system.

B. *Provide Guidance Through A Practices Feedback System*
The project is guided through two feedback processes: (1) the What: agreed specification plan with delivery results clearly tracked in time/cost and resources used/available, and (2) the How: agreed practices that will keep the organization intact and in shape to respond to challenges and changes are tracked through regular surveys and confidential feedback to managers and individuals.

MOBILIZING RESOURCES
The fourth mobilization system is resourcing, a process familiar and essential to engineers.   Every engineering plan demands significant thought to materials and resources, also increasingly evaluating natural resource utilization and environmental considerations.

It is remarkable, however, that many companies have sophisticated systems for information management, for inventory control, but not for the most important fulfillment factor and the only one that can truly improve performance -- people. According to Cooper (1994), human resourcing added during the rework phase can cost in the area of $2000 per hour.   Especially in the partnered organization, human resource systems can produce cost-effective, just-in-time resources.

Three factors are important in achieving just-in-time human resourcing:

+   *HR at the planning table;*
+   *HR processes in the charter agreements*; and

+   *HR in future strategy development.*

In our work with complex organizations with many divisions requiring different kinds of skilled talent, we precede the general strategic meetings with divisional partnering planning sessions in which line managers and HR heads figure out what will be needed.   Some of these sessions have involved experiential as well as cognitive components (Litwin, Bray, Lusk Brooke, 1996) If the HR plan is well-conceived, then chartering can be effective.   As noted before, a multi-partner organization can work together to provide just-in-time resourcing. But commitments have to be made early and be explicitly written into the plan. Unless this kind of provision is agreed, there is a tendency to fall into mistrust and stockpiling.

A final factor in resourcing is knowledge sharing.   Many of the practices discussed above involve organization and multi-organizational learning.   Being explicit (Hamel, Doz, and Prahalad, 1989) about what will be shared and what will be transferred can make partnering a process that both creates new knowledge and safeguards competitive power.

MACRO-RESOURCING -- COOPERATIVE ADVANTAGE
Large-scale projects have a tremendous effect on the regions in which they are located. France put as much money into the upgrade of national rail systems around the country as into the Channel Tunnel.   In Thailand, the engineering and construction boom transformed the craft industry, pulling workers from northern provinces to work in construction.   In fact, one market expert in Thailand commented that the biggest change in Thailand after the boom was not the city, not the industries, and not even the traffic, but the change in family structure: many generations living together in rural areas gave way to a younger generation moving permanently into the city to work in offices. Macro-projects are enormous rocks in local ponds of regional economies; ripples continue to effect change.

Therefore we predict that firms which face the implications head-on by demonstrating and developing contributory policies will have preferred status. There are three areas of contribution:
   • alliance with local firms (and other partners),
   • sustainability and renewal of natural resources, and
   • development of local (and partner) human resources,

as indicated in Exhibit 18.6. Engineering finns seeking macro-projects may wish to demonstrate success in these three areas in order to gain cooperative advantage.

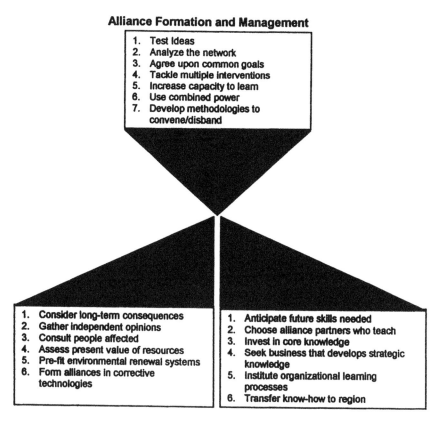

Alliance Formation and Management

1. Test Ideas
2. Analyze the network
3. Agree upon common goals
4. Tackle multiple interventions
5. Increase capacity to learn
6. Use combined power
7. Develop methodologies to convene/disband

1. Consider long-term consequences
2. Gather independent opinions
3. Consult people affected
4. Assess present value of resources
5. Pre-fit environmental renewal systems
6. Form alliances in corrective technologies

1. Anticipate future skills needed
2. Choose alliance partners who teach
3. Invest in core knowledge
4. Seek business that develops strategic knowledge
5. Institute organizational learning processes
6. Transfer know-how to region

Natural Resources Sustainability

Total Human Resources Development

Ex. 18.6 Cooperative Advantage: Essential Strategic Competencies

COOPERATIVE ADVANTAGE

Success on a large-scale is different: the time frame is broader, the investments are greater, the changes are more permanent, and many more people are affected. Therefore, strategy becomes cooperative.

Cooperative positioning in macro-projects begins with the clarity of Shared Purpose. Cooperative agreements and information systems form the system of project infrastructure, while resourcing decisions reflect the impact of large-scale projects on their regional surroundings. But it is in the area of guidance that macro-projects are successful (or not). Utilizing a project-wide metric of practices for performance management, large-scale projects can achieve a level of mobilization that may increase success.

CONCLUSION

Engineering projects have changed dramatically in recent times, largely through regulatory requirements and project size and scope. Both of these trends have increased the importance of partnering.

Especially in projects of macro-scope, management becomes as important as specification. For example, at Stone & Webster, "one day's delay in a completion schedule of a nuclear power plant meant one million dollars in interest alone" (Cameron, 1994). In such projects, what can stop delays when so many partners must coordinate complex delivery?

Partnering requires cooperative actions among many organizations working together simultaneously. Because of the different strategies of the partners, mobilization methodologies that create shared purpose and guidance systems can increase the effectiveness of partnered organizations, resulting in significant success. In macro-engineering projects, organizational and cooperative aspects are of great import; utilizing a shared practices metric for guiding the organization can improve clarity and strengthen performance.

References

Associated General Contractors of America (AGC), (1991) "Partnering: A Concept for Success. Washington, D.C.

Bénard, A. (1997) "Financial Engineering of the Channel Tunnel". In: Davidson, F., Frankel, E., and Meador, C. (eds.), *Macro-Engineering: MIT Brunel Lectures on Global Infrastructure*. Chichester, England: Ellis Horwood, Ltd.

Burns, J.M. (1978) *Leadership*. New York: Harper & Row.

Cameron, E. (1994) "A healthy, productive, balanced world environment means...". In: Davidson, F. and Lusk Brooke, K., *Bringing Macro-Engineering Ideas into Implementation*. Mattapoisett, MA: World Future Society/The Purrington Foundation.

CII, (1991) "In search of partnering excellence," Publ. No. 17-1, Rep., Construction Industry Inst., Austin, Tex.

Cooper, K.G. (1994) "The $2,000 hour: how managers influence project performance through the rework cycle," *IEEE Engineering Management Review*, 22(4) 12-23.

Corbett, G. (1994) "The finance of infrastructure." *Proceedings of the Conference on Global Infrastructure Development*, 1994. Japan: Global Infrastructure Fund Research Foundation.

Corbett, G. (1994) "Lessons learned -- public/private financing -- constructing the model." Brussels: Centre for European Policy Studies.

Cowan, C.E. (1991) "A strategy for partnering in the public sector." In: Chang, L.M. (ed), *Preparing for construction in the 21st century*. New York: ASCE.

Crowley, L.G. and Ariful Karim, M.D. (1995) "Conceptual model of partnering." *J. Mgmt. in Engrg.*, ASCE, 11(5), 33-39.

Davidson, F.P. (1983) *MACRO*. New York: Morrow.

Genta, P.M. and Sokol, N. (1996) "A Microworlds of exploration and development: creating a learning laboratory for the oil and gas industry." *Proceedings of the 1996 International System Dynamics Conference*, 174-177.

Hamel, G., Doz, Y.L. and Prahalad, C.K. (1989) "Collaborate with your competitors -- and win", *Harvard Business Review*, January/February 1989, 133-137.

Kitzinger, U. (1996) "Learning from Major Projects." In: Davidson, F., Frankel, E., and Meador, C. (eds.), *Macro-Engineering: MIT Brunel Lectures on Global Infrastructure*. Chichester, England: Ellis Horwood, Ltd.

Litwin, G.H., Bray, J.J. and Lusk Brooke, K. (1996) *Mobilizing The Organization: Bringing Strategy To Life*. London: Prentice Hall.

Lusk Brooke, K. (1994) *The Momentum of Success: A Study of Construction in the Bangkok Boom*. Mattapoisett: Center for the Study of Success, The Purrington Foundation.

Lusk Brooke, K., Litwin, G.H., and Bray, J.J. (1996) "Cooperative Advantage." Washington: World Future Society International Conference.

Lusk Brooke, K. (1994) "Multi-national, multi-disciplinary projects: chartering for success." *First International Conference, Changing Roles of Contractors in Asia Pacific Rim*, The Chartered Institute of Building,Hong Kong Branch, 43-48.

Lusk Brooke, K. and Litwin, G.H. (1996) "Macro-engineering practices -- what factors are critical for success?" Working Paper #5796, The Center for the Study of Success, The Purrington Foundation, Mattapoisett, Massachusetts.

Oldenburg, C. and van Bruggen, C. (1994) *Large-Scale Projects*. New York: The Monacelli Press.

Saeed, K. (1994) *Development Planning and Policy Design: A System Dynamics Approach*. Aldershot: Avebury.

Senge, P.M. (1990) *The Fifth Discipline: The Art and Practice of the Learning Organization*. New York: Doubleday/Currency.

Weston, D.C. and Gibson, E.G. (1993) "Partnering-project performance in U.S. Army Corps of Engineers." *J. Mgmt. in Engrg.*, ASCE, 9(4), 410-425.

Wilson, R.A., Songer, A.D. and Diekmann, J. (1995) "Partnering: more than a workshop, a catalyst for change." *J. Mgmt. in Engrg.*, ASCE, 11(5), 40-45.

Winch, G.W. (1993) "Consensus building in the planning process: benefits from a hard modeling approach," *System Dynamics Review*, 9 (3), 287-500.

Yennix, J. (1996) *Group Model Building*. London: Wiley.

19

A Civilian-Military Conservation Corps

Janet Caristo-Verrill
President, Macro-Projects International

Macro-projects are the antithesis of the expression, "art for art's sake". Macro-projects are designed to solve macro-problems. They are functional. They serve societal needs and come into being and are successful to the extent that they are integrated into the interdisciplinary fabric of a society and become dynamic structures for sustainable development, social stability, and the flourishing of high technology.

For a macro-project to succeed from conceptualization to realization, a methodology is required. Various methodologies may be combined to achieve a complex societal end. In *How Big and Still Beautiful? Macro-Engineering Revisited*, Meador and Parthe present System Modeling Elements but these elements and literature on macro-engineering in general, do not yield much by the way of methodology and its application. Although there may be tested methodologies for macro-projects other than those which follow, I do not know of any.

Ernst Frankel sets out a most comprehensive methodology in his outline of *Considerations in the Strategic Planning Process.*[1]

The Conflict Management Group of Cambridge (CMG) has designed a dynamic process by which conflict can be resolved within a community by addressing its pragmatic concerns. CMG has had tremendous success in a local Massachusetts community, and Landrum Bolling, CMG's Senior Advisor, has had an even greater success in Bosnia, where he has been working for some time.

My methodology for macro-project formation is based on a model of the Civilian Conservation Corps (CCC) of Franklin Delano Roosevelt's New Deal to which Frank Davidson referred me in 1991 when I was searching for a macro-solution to help order the impending chaos within Russia after the fall of the former Soviet

Union. It was shortly thereafter that I wrote "A Civilian-Military Conservation Corps, A Proposal to Boris Yeltsin and Other Heads of State."[2] The first article on the subject was followed by another in 1996 entitled "Urgent Need for a Conservation Corps."[3]

A CCC provides an organizational principle around which peacemaking and peacekeeping can revolve. It can distract participants from their differences and offer a common objective.

The idea of a Civilian Military Conservation Corps (CMCC) is derived from the tremendous success of the CCC which served to revive the U.S. economy and the morale of its citizens during the period after the Great Depression and before the beginning of the Second World War. A CMCC is a modernized version of the embryonic dreams of two philosophers of the early twentieth century, William James and Eugen Rosenstock-Huessy, which resulted in the most successful program of President Roosevelt's New Deal.

The CCC was introduced in the U.S. in 1933 to alleviate an acute period of despair resulting from hunger and economic collapse. As the program, however, was in all respects a triumph of both political will and the human spirit, it was decided that it would continue as a permanent program for the service, education, and training of the nation's youth and unemployed. It was further decided that the CCC would become socially integrated and expanded to those of the more privileged economic classes through the initiation of Camp William James in Vermont in 1940-1942.

By the time the Second World War came, this slow but more constructive period of economic growth and social stability was drawing to a close. But not before the CCC had enabled the free market to get under way once more by building the foundation, or what is now known as "infrastructure" necessary to a free economy's uninhibited operation. The practical achievements of the CCC and the benefits to the public and private sectors are too numerous to repeat here. Suffice to say, as examples, that CCC volunteers planted thousands of millions of trees, cleared and cleaned millions of channels for flood control, and constructed or maintained hundred of thousands of trails and minor roads in the country, projects that would have been otherwise neglected by the private sector.

The CCC, initially and throughout its accomplishments, enjoyed a 90% bipartisan political approval rating. It saved capitalism at its moment of gravest peril. It is the link or bridge between the New Deal and the "Contract with America" and, indeed, the CCC should be the U.S.'s contract with its trading partners and other members of the international political economy. The CCC is, in all its possible modernized variations, the basic, clear, and grand design or central theme that many seek for a world in transition. The CCC is the "baby" of the New Deal which should now be nourished.

Implementation of a CMCC requires the identification of a country's most fundamental needs and the order in which they are to be satisfied. The needs will determine the structure of the organization and under what auspices it will function. Human, natural, and material resources already available to achieve project completion must be assessed and evaluated, both quantitatively and qualitatively. What is lacking in the way of capital or material investment is to be provided to these self-help, self-directed, intra-country programs by foundations and development banks, and by interested private parties and financial concerns.

The reasons why a conservation corps is so urgent to all societies is the universal exigency for job creation, basic maintenance of a society's physical infrastructure and its social stability, and the planetary need to prevent large migration of populations by increasing their standard of living within their indigenous environments.

A discussion of the CCC as methodology itself would require a great deal of time. What can be comprehended in this short chapter is how the methodology can be given a chance to work or be prevented from working. The Program Director of the military-based Civilian Conservation Corps has nominated me to the Government of Trinidad and Tobago to review their CCC program. The Trinidad and Tobago CCC was initiated by the Executive Branch in 1993 and has trained and educated large numbers of young people who now need to be integrated into the greater economy

For now, however, I will explain briefly how my chosen methodology has applied in failing, failed, and recovering states. I have just returned from Argentina where I hope to introduce the idea to a country where the UN-based concept of "crisis prevention in a functioning but threatened state" (a phrase which may refer to all countries at this critical time in history) may apply. The breakdown in relations between the military and civilian sectors that has occurred during the last decade may be irreparable, but a CCMC could initiate some positive changes in the society as a whole.

Northern Ireland can be viewed as a failing state, not one that has permanently failed or succeeded over the last 1000 years. Despite the longevity of its conflict, it can, nevertheless, be analyzed in contemporary terms to lend itself to a form of CCC with potential to be realized and extend itself across the European Union. Its particular problems of unemployment and a stagnant economy do not relate to fundamental needs which are provided for by the British government. The conflict in Northern Ireland is based in a never-ending culture clash that is founded and perpetuated in institutions which are self-segregating and self-destructive to the growth of a healthy society and economy.

A few courageous individuals have attempted integrated projects, some with more success than others. One, in particular, is Habitat for Humanity International, well-known in the U.S. because of participation in it by former President Jimmy Carter. Habitat's successes are small, however, and I sought to enlarge them by seeking aid from the European Union which offered up to $25 million to projects that were organized for peace and reconciliation. I intended to request paramilitary leaders to recommend 25 each of their best and brightest, with guaranteed immunity, to come out of the cold and see what they could do with a piece of money, power, and responsibility -- hoping eventually to establish these individuals in leadership positions that would encourage others to follow. The idea is that weapons and destructive power would be traded for constructive leadership and management roles in their societies.

I prepared a proposal for the Center of the Study of Conflict in Londonderry and for Habitat, but neither would sign because of the "political implications". While I tried to persuade either or both to sign, the European Union changed its format for giving money for peace and reconciliation and fragmented the amounts so that the projects undertaken became essentially ineffective to anything but reification of the problem. Competition for the money was fierce and counter- productive to the stated end.

While in Belfast I met Ronnie Foreman, who introduced me to a group called Belfast Unlimited, an extraordinary group of community leaders who agreed to back an approach to the paramilitary leaders through their organization. Foreman, who enjoys the utmost confidence of his group, made the necessary contacts and we expected to move ahead quickly when we were abruptly stopped by a violent act at a children's hospital that caused a natural, fearful paralysis in all participants. So, for the time being, we let our plan go by the way.

My experience in Algeria offers an example of a state which was recovering from its ties to the former Soviet Union but recently has become a failed state. It is the perfect example of the principle that "timing is everything". In 1994, at the outset of the worst killings of the last few years, I was invited by the Prime Minister to present my ideas for a CCC to a group of leaders in Algiers. I agreed on the condition that the group I spoke to represented a cross-section of political and professional leaders, including the most extreme. The government agreed to include representatives from all groups, except those who refused to renounce violence as a means to their political ends.

Along with eight Ministers, I conducted a two-day seminar on their existing National Service Program, which was part of their military program, one which the military and government hoped to transfer to the civilian sector. It was a tremendous success. A CCC was formed shortly thereafter, and an Algerian delegation came to Washington and Boston to meet with people at the World Bank and associates of Macro-Projects International.

They returned to Algiers to obtain approval for a housing project that was expected to provide social stability by providing one of the two million houses they needed. This would be done through a CCC program that was designed to train and employ young people such that their numbers would grow by geometric progression, with 8 people training 8 more, etc. The building sites were to be protected by the military. The UN Development Program and the Algerian government agreed to fund stage one and the World Bank stage two of the project.

A "yes" came one day in December, 1994, and the following day an Air France plane was hijacked in Algiers and forced to land in Marseilles. The Algerian airport was off limits to foreign planes for many months thereafter and the government saw little point in going ahead with our plans.

Bosnia is an example of a failed state that could benefit from a CMCC. Little reconstruction has taken place since the Dayton Peace Accords because of a lack of organized action that could run parallel to unending political strife. My recommendation has been that the U.S. military train the Bosnian military to do what it now does for them, and that the Bosnian military do for the civilians what the U.S. military did for the U.S. Civilian Conservation Corps; that is, provide organization, a setting, and discipline for civilian training and project development. The Bosnians could then sustain their own economic development and allow the SFOR troops to reduce in proportion to positive achievement.

A Civilian Conservation Corps in some institutional form can sustain the fundamental needs of a society while at the same time introducing high technology.

For example, what was planned for Algeria was to introduce and train the young and unemployed in computer skills as they relate to construction and design. High-tech skills would then spill over into other areas of the economy, fundamental needs would be established, and high technology could flourish. Many possibilities for the introduction of social stability and the achievement of pragmatic ends exist on the African continent, in particular, in countries such as Nigeria, Rwanda, Zimbabwe and others. What is needed is a Macro-Projects Center through which information from one program to another can flow and from which self-help programs can be designed.

NOTES

1. *Newsletter of the American Society for Macro-Engineering.* Third Quarter, 1997.

2. *Interdisciplinary Science Reviews*, Vol. 17. No. 2., 1992.

3. *Interdisciplinary Science Reviews*, Vol. 21, No. 3, 1996.

Section Five

FINANCIAL AND LEGAL ISSUES

20

Prime Contracting:
Is it Profit with Peril?

David W. Stupples
Project Management Practice
PA Consulting Group, London

INTRODUCTION

In many developing countries, the rate of economic growth is significantly ahead of infrastructure investment and, because of other priorities for government spending, there is a need to attract private investment for new infrastructure. The build-operate-transfer (BOT) project arrangements would seem to be the way forward, offering a "buy now and pay as you use" concept. It is becoming the practice to use the prime contracting instrument for BOT arrangements, with one contractor taking full responsibility as a business undertaking. Owing to the complex organizations involved, the many conflicting roles that a government must play in the project, and the unpredictable nature of domestic issues, prime contracting BOT projects is not without its perils. By implementing specific management measures, a prime contractor will be able to contain these perils, but will need to convince the government and other players and stakeholders involved that the additional management technique is a value-added service.

Infrastructure projects are inherently large, expensive, complex, and sometimes speculative, and most face implementation difficulties owing to their effect on the socio-economic and natural environments. Furthermore, infrastructure projects are often implemented over extended time periods and are therefore exposed to political uncertainty and possible adverse public reaction. Hitherto, infrastructure projects in the developing economies have been directly financed from public funds, with governments taking the responsibility for design and operation and a contractor being selected to construct the facility. A published World Bank special report[1] confirms earlier estimates by World Bank economists that between $1,300 billion and $1,500 billion will need to be spent by Asian and Pacific Rim countries on infrastructure in the next decade in order for it to be capable of supporting the economic growth. This figure more than doubles when infrastructure funding for Eastern European and Latin American countries are included.

The point has now been reached at which governments are no longer able to manage or fund the majority of new infrastructure projects and are therefore encouraging private sector participation through the BOT project concept.[2] This arrangement involves the selection of a prime contractor who will take responsibility and risk through a concession agreement for designing, building, financing and operating the project for a determined period of time. Essentially, BOT projects are complex business arrangements in which investors in the project will seek a fair return on their investment.

Most governments accept that the BOT project arrangement is the way forward, particularly if they wish to avoid outright privatization, but there are pitfalls associated with the complex project structures involved, the many and varied roles held by government, and the instability of conditions for investment. The key to success is to define at the outset how the customer, the commercial aspects, the construction, the in-country issues and the interfaces that exist between them will be managed, and to design an appropriate project to deliver the infrastructure.

PRIME CONTRACTING BOT PROJECTS SEEMS TO BE THE WAY FORWARD

Unlike those in Latin America, governments in Asia and the Pacific Rim (with the possible exception of Malaysia) have rejected privatization of their infrastructure, opting instead for BOT arrangements under the cover of a prime contract. There is widespread agreement that the BOT concept preserves national ownership while at the same time accelerating the provision of much-required infrastructure facilities. The BOT concept[3] is shown in Exhibit 20.1. BOT projects, build-own-operate-transfer (BOOT) projects, and build-own-operate (BOO) projects are all derivatives of the same basic premise and relate to specific business propositions.

- Design
- Manage implementation
- Procure facility
- Build facility
- Finance facility

- Obtain concession to operate
- Arrange franchises
- Manage and operate facility
- Maintain facility
- Receive payments

- Plan and manage
- Transfer facility to owner

Build

Operate   Transfer

Ex. 20.1   **The "build-operate-transfer" arrangement is a "buy now and pay as you use" concept**

Limiting and spreading risk is an important element of government-sponsored BOT projects, but this can lead to a large number of stakeholders being involved (often as many as 20 in £500 – £1,000 million infrastructure projects). Equity is essential, the amount being dependent upon the market's perception of the scale of the risk; with unlimited non-recourse financing, it is often as high as 25%. Apart from equity financing, other conventional sources of finance include subordinate financing and junior debt (preference shares, warrants, convertible bonds), senior debt (commercial medium- to long-term bank loans arranged through banks), and leasing (where the owner can claim capital allowance for equipment, and offset it against corporate taxation). A typical BOT project structure[4] is shown in Exhibit 20.2.

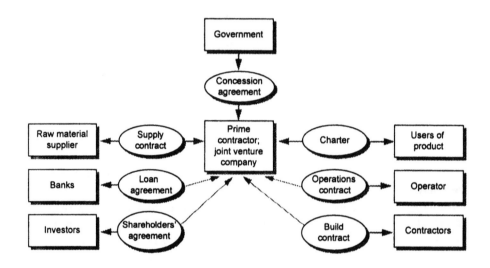

Ex. 20.2.   The BOT project structure reflects a business relationship

The structure involves the creation of a special-purpose joint venture company established as a prime contractor in which the new company formation, operator, and banks have a share. This joint venture company, sometimes referred to as a concession company, borrows money to fund the construction on the security of the revenue that the lending banks believe will be generated by the project. All financial obligations must be serviced within the life of the concession. The project is approached in a similar way to limited recourse project financing in which the risks are isolated and allocated to those most qualified to bear them. At the end of the concession period, the project and the right to operate it reverts to the government, which may then choose either to grant a new concession or to operate the facility itself.

In order to distribute risk, governments look for a debt to equity ratio of 30% to 70%, respectively, with domestic stakeholders being encouraged to take a significant share in both elements. However, it is theoretically possible to finance a BOT project entirely from debt without there being any requirement for equity. This means that what would otherwise be equity risk is channelled through one or more of the contracts.   The $1.6 billion Bangkok Transit System (BTSC) will be almost entirely financed through private funding, with around half being provided by domestic stakeholders.[5]   The $17.5 billion controversial Three Gorges Dam (Yangtze River) project will be financed almost entirely by China's State Development Bank (SDB),[6] but it will raise funds for this and other strategic infrastructure projects on the international markets.   The $5 billion Thai Digital Telephone System project, although not strictly a BOT project, is being procured through a prime contract with TelecomAsia (a joint venture company);   finance for this project has a typical debt/equity split, with 50% of the funds being provided from domestic markets.

Most governments in the developing economies now accept that the BOT concept is an effective means of providing new infrastructure on a "buy now and pay as you use" basis, often referred to as privatized infrastructure.   Through the combination of the BOT concept and the prime contracting instrument, governments are able to lay off risk, yet, through their position as project owner, still exert considerable influence over affairs.   A prime contract encapsulates the concession agreement in a legal framework for project direction and discipline.   The contained concession agreement enables both contractors and financial institutions to achieve a fair return on investment and therefore, if set up properly, a win-win situation is established.

THERE ARE PERILS IN PRIME CONTRACTING BOT PROJECTS

The perfect marriage between a government and a prime contractor for the delivery of new infrastructure through a BOT arrangement has always been the goal, but is seldom achieved in reality.   The number of players and stakeholders involved makes for a complex organization with interfaces across political, company and geographic boundaries;   the many and varied roles of the customer often introduces uncertainty into the contractual relationship;   invariably, the domestic situation is volatile.   It is important to understand the nature of the perils involved in order that a good prime contract management structure can be established.

The Number of Players Involved Makes Complex Project Organizations

The involvement of many players and stakeholders in a BOT project (see Exhibit 20.3) gives rise to a complex organization that a prime contractor must manage if the project is to be maintained on a stable footing.

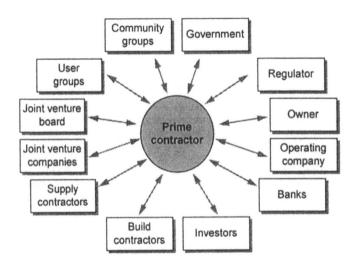

Ex. 20.3   The prime contractor has a complex organization and an array of interests to manage

The different interests involved tend to exert pressures that are incompatible with a group sharing a common cause, especially when differing agendas emerge from each crisis occurring within the project or within its direct sphere of influence. In most instances, these differing agendas arise from domestic economic pressures or those emanating from project delays or unexpected and cost-sensitive technical difficulties. BOT projects are particularly sensitive to changes in risk, with participating players and stakeholders responding differently, sometimes emotionally and irrationally, in accordance with their perceived interest. For large BOT undertakings, this has the effect of pressure ripples in the organization, thus ensuring that it is rarely in a stable state.

The $14 billion Anglo-French Channel Tunnel undertaking epitomized the BOT organizational problem (although, strictly speaking, it followed a BOOT arrangement). Following an initial honeymoon period, the concession contractor, Eurotunnel, was involved in a continuous exercise of appeasement and nerve-calming as successive project difficulties sent shock waves into the organization.[7] Most of the shock waves were typical of large projects but became amplified and distorted owing to the diversity of interests and the large investment involved. In its "own-operate" role, Eurotunnel is still entrenched in organizational difficulties as project economic forecasts fail to meet expectations.

Governments can be Complicated Customers

A government as a customer presents any major infrastructure project with difficulties owing to the many roles it must play. In any democracy, a government is accountable to the electorate for the overall economy, the quality of life of the population, fiscal policies, and expenditure. For infrastructure projects, these roles include that of sponsor, regulator, champion and owner. However, owing to the non-monolithic structure of government, these roles are distributed through a number of departments, each with its own responsibility. The result is that the prime contractor does not see a customer with a common purpose and aim:

- A government will sponsor an infrastructure project either as part of an election pledge or as part of wider government policy. Through its elected members, central or local government can also bring pressure to bear on any infrastructure project if it serves a political end. This pressure may be with or without the prime contractor's knowledge and it may not help the project to fulfil its defined objectives.

- In its role of regulator, a government will ensure that new infrastructure is in line with existing and planned legislation and any internationally agreed standards, and that the public is protected from any resulting monopoly operation. In many instances, especially with telecommuni-cations infrastructure projects, new legislation lags the build project, with the consequence that the regulator slows the project, thus making an impact on the business case. Often, the sponsor of the project is powerless to overrule a regulator's directive.

- A government or an influential department of government will champion a politically exciting project, with the result that the government machine is behind the initiative. This moral backing is needed where private investment is required for high-risk projects, particularly for countries with a history of political instability. A government's moral backing will flood and ebb, reflecting either the fortunes of the party in power or the economic climate.

- The government acts as an owner through one of its many departments. The management of the project, through that department, may not necessarily be in tune with its political masters. Furthermore, the decision chains are often long and bureaucratic and their members reticent to commit on issues that might affect government policy without deference to the political leaders.

Ideally, the government as the eventual owner is the single point of contact for the prime contractor and this link is the conduit for all government and prime contractor decisions. However, the prime contractor needs to establish links to the sponsor, the regulator, and the project champions if political pitfalls are to be avoided.

Domestic Issues Can Make for a Volatile Market

Infrastructure projects, by their nature, have an impact on the communities that they serve[8] and, as a result, socio-economic, political and environmental issues become important. The impact of these issues is not constant, but varies markedly as a result of changes in the domestic climate:

- *Political instability:* The concession period for the project often exceeds the term of government office and the project may well be subject to a sudden change in policy and direction, often without compensation.

- *Underdeveloped legal framework:* Many developing countries do not as yet have a legal framework that can support the BOT concept and its associated concession agreements supported by the prime contracting framework. As a result, it is difficult to set up international consortia because the legal system does not offer the degree of protection required for financial investment, especially when a significant equity stake is being made.

- *Underdeveloped monetary framework:* Foreign investors require repayment in a tradeable currency throughout the concessionary period in order that they can readily realize their investment in the open market. In many developing countries, governments manipulate foreign currency exchange to suit immediate macroeconomic goals, often to the detriment of existing commercial agreements.

- *Pressure groups:* These groups include environmentalists, local governments, trades unions, and community interest groups. Pressure groups can cause disruption to an infrastructure project, putting continuing financial support in jeopardy. In countries in which the government is weak, these groups can cause major change or even cancellation of the project.

Domestic issues cannot be controlled, but their impacts can be anticipated and plans made to contain their impacts or mitigate their effects accordingly. Lenders and investors have developed both a risk assessment and risk rating for each country that will raise potential issues in the early stages of BOT project planning.

THE PRIME CONTRACTOR'S ROLE IS TO DEAL WITH THE FOUR KEY DOMAINS OF THE PROJECT

The prime contractor's role in international BOT infrastructure projects is to deal with the four key domains of the project – the customer, the commercial aspects, the construction (or build), and the domestic issues. The latter can involve international considerations where cross-border issues are involved – for example, new telecommunication systems with international links, and inter-country highways. The management arrangement and focus of the prime contract, shown in Exhibit 20.4 reflects the BOT project being managed as a full business undertaking. The domain overlaps represent the project that must be designed for a particular situation.

The overlap of the domains denotes that a specific implementation of the domains is required for each prime contract. This specific implementation has to be designed to meet the needs of a particular prime-contracting undertaking. Designing the project is about assembling the management structures, the control and reporting structures, the commercial structures, the design procedures and methods, and introducing an acceptable project culture and team working. Every prime contract project is different, and therefore each requires careful design consideration if it is to be successful.

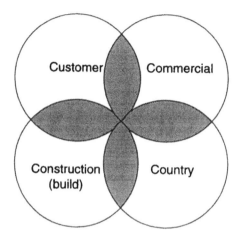

Ex. 20.4   The prime contractor's role is to manage the four domains of the BOT project

Customer management represents the main interface with government
Customer management maintains and controls the interface between government and the BOT project. Its two primary roles are the information conduit with the customer's disparate organization, and the project's main contact with the regulator. The latter will entail ensuring that the BOT undertaking is in accordance with national and international standards and regulations, and that the facilities provided are in the public's interest. A secondary, but important, role is to establish a rapport with the government individuals who sponsor and champion the project in order that political support can be maintained. This secondary role becomes more important during times of political and project difficulty, at which point political lobbying from in-country representatives may be required.

Commercial Management is Responsible for Running the BOT Project as a Business
Commercial management is responsible for establishing and maintaining the BOT undertaking as a business. The activities associated with commercial management include:

- Project economic monitoring
- Debt management
- Investment management
- Operations management
- Management of revenues
- Management of foreign exchange issues
- Addressing fiscal issues and legal matters
- Management and cost accounting
- Joint venture management.

Commercial management has been poorly addressed on a large number of international BOT projects, with the result that the projects have not realized their full potential, or have experienced difficulties that could have been avoided. Many have failed to appreciate that BOT projects are complex business ventures, sometimes undertaken in a medium- to high-risk environment. At all times, the project is open to domestic socio-economic and environmental influences and the unwarranted interference of the players and stakeholders involved. Good commercial management will contain the perils in this area.

Construction Management is the Design Authority
The prime contractor is responsible to the owner for ensuring that the BOT project meets its requirements definition. To achieve this, the prime contractor retains authority for the design and associated change control, and oversees system integration and commissioning (particularly important for telecommuni-cations or computer system infrastructures). Normally, the prime contractor contracts out the construction or build of the infrastructure facility to one or more companies. The construction or build management acts as the single point of contact for all build contracts in order to maintain its integrity as design authority.

Country Management Looks After the Domestic Issues
The primary role of the prime contractor in this domain is to address the domestic and social aspects of the project. On many infrastructure projects, dealing with pressure groups and local politics represents a full-time role and, if undertaken properly, should involve close work with the local community, particularly pressure groups, to address their concerns before they begin to have an impact on the project's economics. It is now generally accepted that, for this work, local staff are preferred to expatriates as they understand more readily the local culture, traditions and language. Furthermore, it adds to the belief that the project is a national venture, thus increasing public acceptability.

The Domain Overlap Represents the Design of the Project
The design of a project should first be addressed during project definition where a full understanding of project size, complexity, technical uncertainty, schedule duration and urgency, physical and social environments, business situation, and government and politics is realized. The domains can then be effectively focused on the needs of a particular project situation using the design process shown in Exhibit 20.5.

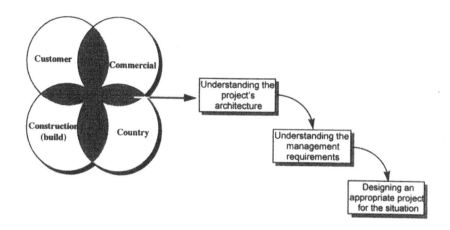

Ex. 20.5 the design of the design is in the overlap

The aim of the project design will be to achieve an appropriate structure to drive the project to success. The design process comprises three elements:

- *Understanding the project's architecture* is about identifying all of the project components and their interactions, and forming the best structure for a particular situation. From this structure, it is possible to develop the optimal configuration for the prime contractor to operate.

- *Understanding the management requirements for the project* involves recognizing the control and reporting needs of all project processes and the required mechanisms for controlling project interfaces, whether they be dynamic or static. Dynamic interfaces connect the project lifecycle stages whereas static interfaces pertain to the project's existence within its environment.

- *Designing an appropriate project for the situation* is about identifying an appropriate organization, defining roles and responsibilities, establishing project processes, specifying operations frameworks and selecting appropriate methods and systems.

Project managers should be aware that the environment surrounding a project is never stable and that often a project's design will need to change to reflect a changing environment. For large infrastructure projects, the project manager will need to establish sensors and a responsive management culture to ensure that the project is always set up for success.

The provision of international infrastructure through build-operate-transfer arrangements must be considered to be profit with peril owing to the organizational complexities involved, the diverse roles of government, and the undetermined nature of domestic politics and issues. These perils can be contained if a prime contractor defines, at the outset, how the customer, the commercial aspects, the construction (that is, the build), the in-country issues, and the interfaces that exist between them are to be managed. However, a prime contractor must first convince governments and other stakeholders and players that the management service offered is essential to the success of the BOT undertaking in both its project form and the subsequent business form, and that all involved must use this service to influence the project in a coordinated and agreed manner.

25

NOTES

1. Kohli, H. *Infrastructure development in East Asia and the Pacific*, World Bank publication, September 1995.

2. Montagnon, P. "Asian projects lagging", *Financial Times*, 14 September 1995.

3. Haley, G., "Private finance for transportation and infrastructure projects: a view", *International Journal of Project Management*, Vol. 10, No. 2, May 1992.

4. Ibid.

5. Bardacke, T. "Rail scheme faces tough funding choice," *Financial Times*, 7 September 1995.

6. Montagnon, P and Walker, T. "Providing funds for China's infrastructure," *Financial Times*, 12 May 1995.

7. Taylor, A. "Pounds, pioneers and more pounds," *Financial Times*, 3 May 1994.

8. Tam, C. "Features of power industries in South East Asia; Study of Build-Operate-Transfer power project in China," *International Journal of Project Management*, Vol. 13, No. 5, October 1995, pp. 303-311.

21

Managing Price Through Knowledge Networking

Sean Gadman
Vice President, Marketing & Innovation
Pricetrak (USA) Inc.

Wally Johnson
Chairman and CEO
Pricetrak (UK) Ltd.

INTRODUCTION

Macro-engineering projects by definition tend to be long-term, high-cost and commercially complex. The key issue is managing price over time. The longer the project period, the more challenging management becomes. This is due, in large part, to the simple fact that the world within which these projects operate is growing increasingly complex and ambiguous. These are ideal conditions for speculative suppliers and contractors to capitalize on large scale projects by withholding and distorting price information.

This chapter proposes a new way of managing the price of large-scale engineering projects, a way that surfaces covert pricing structures through networking the knowledge of purchasing professionals. With this knowledge, planners can make informed and real-time decisions on the price of bought-in goods and raw materials, thereby informing the procurement decision-making process.

THE PROBLEM

Macro-engineering projects share similar characteristics, that is, they extend over long periods of time, are very high cost, and are commercially complex. A key issue is managing price over time -- the longer the project period the more difficult price management becomes. Anything beyond two years is a challenge as the following example of the Eurotunnel shows:

1981 September:	Thatcher and Mitterand announce studies for a fixed channel link.
1986, March:	British and French sign a concession with Channel Tunnel Group and Frenche-Manche to last 55 years from the completion of the legislation.
September:	Equity I: Consortium raises 46 million pounds sterling
October:	Equity II: Eurotunnel raises 206 million pounds sterling through private placing
1987, February:	Sir Alastair Morton arrives as co-chairman.
April:	First public row erupts between Morton and contractors.
May:	European Investment Bank lends 1 billion pounds sterling
December:	Construction begins
1988, August:	Eurotunnel warns of delays.
1989, July:	Eurotunnel seeks fresh loans of 1 billion pounds sterling as costs rise.
October:	TransManche Link warns costs will rise to 8 billion pounds sterling, which Morton dismisses as nonsense.
1990, October:	Agreement signed for extra 1.8 billion pounds sterling.
1992, February:	Eurotunnel opening delayed from June 1993 to end of summer.
1993, April:	Row with TML erupts over 1 billion pound sterling of bills. Eurotunnel warns opening could be delayed by more than a year.
October:	Eurotunnel says it wants an extra 1 billion pounds sterling to keep the tunnel in operation after the opening
1994, May:	Tunnel opened by the Queen and Mitterand. Also unveiled is a 925 million pound sterling loss, one of the biggest in corporate history.

Long-term projects often suffer from late specification changes as designers and project managers try to incorporate later development into the design or rectify newly identified faults. It is fertile territory for cost overruns, sometimes on a horrific scale. Less frequently, contractors run into difficulties; some, although not many, have been bankrupted or nearly so. For example, in the case of Eurotunnel, major debt restructuring is now being considered in order to repay debts within the concession years.

Managing price over time is a serious challenge because fixed pricing is a gamble for all parties. Contractors, unable to go back to back to their suppliers, can find input prices rising beyond reasonable expectations, particularly in the case of commodities but also of manufactured goods and uncertain labour situations. For the project owner, fixed pricing will include a premium based on the contractor's estimate of price changes over the period of the contract to which is added another premium as insurance against uncertainties. For a four-, five-, or six-year project period these premiums can be substantial. Contractors win projects on their willingness to gamble.

Conventional pricing arrangements for these projects involve three pricing elements:

- formulae (intended to be a working cost model of the project)
- indices
- application rules that are used to change the formulae as circumstances change. (A decision to double the size of the project or extend the length of time for delivery leads to calls from one side or the other for a formula change.) These are often covered by rules but not always, and it is not possible to forecast all the possible changes in contract parameters.

Experience has shown that all three of these factors can soon become unrealistic. The authors know of one case where contracts for supplies of oil from a refinery to an adjacent power station led to a situation in which soon after the contract started the maximum price provision in the formulae fell below the minimum price. Additionally, the chosen indices may not be appropriate to the model or the indices may themselves not truly represent what is happening in the market.

Formulae tend to underestimate the fixed-cost proportion to the advantage of the contractor. They can be exceedingly complex, yet still fall short of being accurate cost models. The timing of changes in formulae and indices can be critical and call for understanding of transactions and manufacturing dates. Initiatives in the development of pricing mechanisms can be a one sided affair. Manufacturers with a less than perfect understanding of their own cost structures have specific advantages over the customers whose access to information is even more limited. Uncertainties are a winning card in the contractor's pack. Some contractors intentionally under-price at the bid and negotiation stage with expectations of later recovery if they get the contract through the operation of price variation provisions. The process is a game between unequals.

THE CHALLENGE

The root cause of these inequalities is the difference between overt and covert price levels. In practice, formal price lists and producer and exchange prices are seldom those at which the products in question change hands. They may offer bases for driving price, but list price relativity is at best crude and at worst totally misleading. It is, in any case, a high "fog" area as manufacturers and suppliers use multiple price lists to obscure their real pricing policies. Withholding information is a first-class strategy for increasing profits. As we all know, information is knowledge and knowledge is power.

In retailing, access to selling price information is available (at a price) from market research agencies constantly sampling prices as a commercial service. In industrial, commercial, and institutional transactions, price is resolved by private treaty after negotiation or bid processes, and little quantitative information is generally available about the true levels or behaviour of prices. At Pricetrak Inc., our own data suggests that real prices in these sectors are far less volatile than would be inferred from the behaviour of formal pricing structures.

Prices do not reach the same highs and lows, and the incidence of real price increases lags behind list price increases as ongoing contracts give established customers some breathing space before impacting. The reverse happens on falling markets. To complicate the issue further, for international transactions (and macro-engineering projects often are) the question of exchange parities involving more than one currency adds a further dimension to the gamble.

When the contractor is part of the same organization as her or his suppliers, there can be further obfuscation. Vertical integration in the engineering and construction industries is not uncommon and transfer pricing provides opportunities to exploit cost-based pricing mechanisms. While recognizing limitations in the availability of data relating to real price behaviour in the commercial, industrial, and institutional sectors, the authors believe that the future lies in market-related pricing for long-term contractual relationships.

MARKET RELATED PRICING

One key element in effective purchasing is market-related price information. Collected and organized, it can be used to make comparisons not only about who pays how much, but for what, and by whom. Such price monitoring equips purchasing specialists to deal more effectively in markets, as well as prompting wider questions about specifications, supplier performance, stockholding, and internal procedures and systems. More importantly, by providing some measure of comparative performance such price monitoring enables purchasing professionals to innovate when it comes to managing prices.

Using computerized price monitoring systems like those at Pricetrak Inc. it is possible to establish a network of purchasing specialists who share their knowledge about market conditions. They do this by entering the price paid for a particular item, the quantity, and supplier into a common database. These data are then processed into three averages:

1) An arithmetic average
2) A weighted average

3) A lower quartile average showing the average price paid for the lowest 25% of purchases.

Particular trends are highlighted and buyers in the knowledge network compare their purchasing performance with each other.   Where the community exists within a specific project, purchasing specialists can compare notes and develop informed strategies prior to entering into negotiations with suppliers.   Such knowledge networks are ideally suited to the kind of volatile environments that exist in large-scale engineering projects where sizable purchasing activity takes place across a highly dispersed community.   Informed and intelligent buying practices based on market price movements contribute significantly to a projects cost predictability.

Drawing on the growing trend in leading international business toward leveraging their intellectual capital as a key asset, computer-based price management systems provide a refreshingly new perspective on the role and value of the purchasing function. They enable purchasing specialists within a project to identify, structure, capture, store, retrieve, and leverage the living wisdom of all specialists. It values and capitalizes on the wealth of knowledge carried by purchasing specialists as they interact with one another both inside the project and anonymously, across companies.   Most important of all, these systems provide the foundation for more informed and innovative dialogues between purchasing people, their customers, their customer's customer and their suppliers.

SUMMARY

A new medium of human communications is emerging, one that may prove to surpass all previous revolutions in its impact on our economic and social life. Microcomputers and advanced electronic communication systems are enabling a new economy based on the networking of human knowledge. In this knowledge- based economy, enterprises will create wealth by applying knowledge in entirely new and innovative ways. Their success will be determined more in terms of their ability to acquire, generate, distribute and apply knowledge than in terms of their hard assets such as trucks, assembly lines and inventory.

Nowhere is this more in evidence than in the complex world of purchasing and supply. The value of procured goods and services has now risen to nearly 70% of the cost base in large-scale engineering projects. There can be no doubt that a well-constructed, concentrated supply base contributes significantly to better price management and ultimately to lower costs.

22

Concepts of Eminent Domain

Jon N. Santemma
Co-Chairman, Condemnation & Tax Certiorari Committee
New York State Bar Association

In order to bring macro-engineering projects to fruition, the sponsor of the project will require real estate -- probably huge quantities of real estate -- possibly located in different jurisdictions. The issue will be how to obtain title to the property which is essential to the project.

In this chapter, we will consider some of the difficulties in obtaining the necessary sites through purchase, and then analyze the concepts of eminent domain or compulsory purchase.

PRIVATE ACQUISITION

In the book, *Macro-Engineering and the Future*[1] the editors point out that a key work in the discussion of large projects is "community". They note"

> ... Even a nation committed to individualism and free competition must retain a sense of common purpose and destiny if it is to deal effectively with questions of survival. There is an irreducible minimum of common effort and common obligation without which there can be no community, no law, and therefore no collective future ....[2]

The tensions which the editors describe, between free enterprise and the concept of the community - are unhappily reflected in the "NIMBY" philosophies opposing public projects. Everyone is in favor of all of the good projects which would provide for the common good: hospitals, research, roads, telecommunications, transportation, etc. etc. -- but "Not In My Back Yard."

Take, for example, a proposal to install a high-speed train right of way from Boston to New York, which requires an absolutely straight right of way. The proposed route would traverse the Commonwealth of Massachusetts and the States of Rhode Island, Connecticut, and New York. Each of these sovereign states and the Commonwealth have created various townships, villages, hamlets, cities, counties, and other governmental units, each one of which has regulatory powers over what happens within its borders and each one of which has, in the past two or three decades, become increasingly concerned about the environment and the impact of major projects on the local environment and the local economy. Even things which objective perspectives deem "minor" become *cause celebre* in local neighborhoods.

The increasing concern on the part of the general population for environmental and quality of life issues, coupled with the prevalence of the NIMBY philosophy, makes it difficult, in many jurisdictions, for anyone to proceed expeditiously to put an addition on their home, much less inject a macro-engineering project. The multitude of state and local agencies that would be involved in every aspect of a Boston to New York Mag-Lev train, for example, would not only absorb decades of time in public hearings debate and analyses, but the cost of such applications and hearings would be beyond comprehension -- and the likelihood of receiving required approvals from the totality of state, county, town, city, village, hamlet, special districts, and jurisdictions along the route approaches zero, even if each of the owners of affected property wished to sell at reasonable price.

THE POWER OF EMINENT DOMAIN

National, state, county, or local government, however, is able to accomplish projects wherever and whenever that government determines that the project is to be undertaken. A good example is the interstate highway network which was conceived as a defense necessity following the Second World War and which led to the entire continental United States being criss-crossed with a grid of interstate highways created under the aegis of the federal government and constructed either directly with federal funds (a FISH project - federal interstate highway) or by the states through the use of federal grants (a FASH project - federally assisted state highway). In either instance, however, the power to acquire sites is an inherent hallmark of a sovereign government.

The power to take property from private ownership and use it for the common good is an attribute which is among the most powerful of the sovereign entity's powers since it may be unilaterally exercised by the sovereign simply because the sovereign determines that the appropriation or acquisition is in the best interests of the common good. The taking will then proceed over the most vehement or vocal opposition, and the courts traditionally consider the public nature or the taking and the location of the take to be legislative functions which are not reviewable if they are for a public use. The term "public use" has been very liberally interpreted.[3]

The use of the term "eminent domain" is reported as first being utilized by Grotius and there are earlier references, even Biblical circumstances, which can be analyzed in terms of eminent domain (see, for example, 1 Kings 21, verse 2). Whether viewed as the right of the sovereign to exercise its paramount title to all property within the realm, or more as a common law attribute of the sovereign to take what is necessary to permit it to do sovereignly things, the fact of the matter is that the power is present in each state as successor to the sovereign powers of the king.

In England, the concept has become known as "compulsory purchase". Aside from the power of the Crown to take that which is necessary for the defense of the Realm, the concept is first identified with the need for allocating uncultivated land in the thirteenth century as common land, and later in reallocating farmland ownership as feudalism was replaced by modern agricultural practices in the seventeenth and eighteenth century.[4]

THE STATES AND THE FEDERAL GOVERNMENT

As one of the most ancient of sovereign prerogatives, the power to take private property for common public good has survived in each state as an attribute of the sovereign.

As a result of their status as sovereigns, successors to sovereigns or agents of sovereigns, states in the U.S. enjoy the power of eminent domain. That is, they can determine that it is in the best interests of their municipality to acquire certain property and devote it to a public use. The states are obligated to follow certain procedures in each instance, and they are constitutionally mandated to pay just compensation for that which is taken by virtue of the Fourteenth Amendment to the U.S. Constitution as well as by their own state constitutions.

The federal Constitution conditions the federal government's right to exercise the power of eminent domain upon the payment of "just compensation" for that which is taken. Some states may be more liberal and not limit compensation to that which was actually taken, but also pay for that which is demonstrably damaged. The Fifth Amendment to the U.S. Constitution merely prevents the taking of private property without just compensation.

In our example of the Mag-Lev Boston to New York train, assume that each of the states of Massachusetts, Rhode Island, Connecticut, and New York determined that it was appropriate for them to participate in the project and to own the Mag-Lev right of way in their respective states. Each of them could begin, through a series of proceedings started in each state, to acquire the right of way as necessary within the four states.

Each of the states has a procedure for acquiring property, each has standards for considering the environmental impact of the taking, and each has procedures that enable the state to acquire corridors of land for the purpose of acquiring sites for transportation purposes.

However, it is conceivable that the states in the middle of such right of way might not share the enthusiasm for the project to the same extent that those located in the terminal states would enjoy. An adverse determination by any one of the four states would, of course, prevent the acquisition to take place in that state and would, therefore, doom the entire project.

A FEDERAL TAKING

The federal government could, if it wished, acquire a right of way from Boston to New York, in a relatively simple procedure which involves the same steps as are necessary to acquire land for a federal park or to declare a federal national seashore or establish federal military base or other facility. The Fifth Amendment to the

Constitution gives to the federal government the power of eminent domain, but specifically conditions the exercise of the power on the payment of just compensation. When the federal government exercises those rights, the amount of compensation is determined in the federal court by a judge and jury, which is a cumbersome procedure, and the federal government becomes the owner of the project.

PRIVATE USE OF A PUBLIC POWER

A current macro-engineering situation is analogous to the problems confronting the railroad companies more than a century ago as they proceeded to establish new routes in unprecedented expansions throughout the United States. Similar as well are the problems faced by the utilities which required distribution facilities to enable the electricity, telephone, and telegraph services to be provided throughout the continental United States. All of these had to breach the private ownership of an entire country in order in order to link the east and west coasts, the Canadian and Mexican borders, and the parts in between.

The issue is also analogous to the realignment of land titles in feudal England to create areas of "commons", and then to use the concept of compulsory purchase to realign the agricultural community through enclosures. At the height of the enclosure activity in seventeenth and eighteenth century England, each project was sponsored by a promoter who was the most prominent of the land owners seeking the enclosure of the open land. The promoters would appoint commissioners to hear and determine all claims, make an award which would constitute a comprehensive plan for all of the property, establish roads, schools, warehouses, and other particular facilities, and then apportion the lands among the various claimants. The promotor would then apply for Parliament's confirmation.[5]

These twelfth century concepts of an involuntary purchase by the sovereign of quantities of land deemed to be uneconomic in their then current usage, and consolidating or redefining use and ownership for resale, are exactly the philosophy underpinning modern urban renewal acquisitions. In modern urban renewal, the federal government provides fundings to independent local agencies charged with the responsibility of acquiring properties in an economically blighted area and providing for a reuse of the area so as to restore economic vibrancy.

The theory is to provide the financial wherewithal to a locally sponsored independent agency created by the government with the power of eminent domain so to permit localities to solve their own economic problems with local people and federal monies.

Similarly, railroads, water companies, electric companies, telephone companies, and telegraph companies were given status as transportation companies. That class of corporation received the power of eminent domain as a result of their classification as transportation corporations, and they were able to acquire the sites necessary to provide railroad, water, electricity, telephone, telegraph and other services to the residents of a particular area.

Even today, if such company is to expand its service in a particular area, it may do so through acquiring the necessary rights of way in eminent domain. Its only obligation is to pay for the value of that which it took, all as determined in just compensation.

Whether or not any macro-engineering project requires federal money it will, by its nature, require the approval and support of the government. In moving a project forward, it is essential that the federal government in particular give to the sponsoring organization the power of eminent domain just as that power has been extended to numerous railroad corporations, utility corporations, public authorities, urban renewal agencies, and other entities in which the public and private sectors interface in order to produce a desired result for a community, whether that community is as small as a neighborhood or as large as a transportation corridor.

There is a significant difference in the amount of government control in transportation companies and in urban renewal agencies. Urban renewal agencies are essentially funded by the federal government, and while the local agency is supposedly autonomous, virtually every phase of its operation which spends the federal money is analyzed, recorded, and controlled.

The transportation corporations are not similarly treated. They pay for that which they acquire through their own treasuries, recouping the cost through their regular fees and charges, regulated or not.

JUST COMPENSATION
The constitutions of the federal and state governments do not specify the measure of damages to be paid, but speak in terms of just compensation.

In the vast majority of cases, just compensation equates to market value, and in most states only to the market value of the real estate damages generated. In most states, the owner of a business operated on a condemned piece of real property is entitled to no compensation on the theory that the business is not that which was taken -- only real property was taken, and no compensation need be paid in the absence of a taking.

Generally, real estate damages are determined from market value considerations.

With respect to the costs of site acquisition, the condemnor is obligated to pay just compensation as measured under a variety of pertinent valuation principles. For example, if a railroad station site is to be acquired upon which a station can be built, the value of the property that is acquired would be fixed in accordance with the square foot value of comparable sites in the community. If the property has income-producing property on it already, or single family or multi-residence housing, each of the separate uses and entities is valued under traditional income analyses or market analyses and an award made to the owner of the property that is acquired.

With respect to the right of way between the two Mag-Lev stations from Boston to New York, the cost in land acquisition varies in direct proportion to that which is taken. Obviously, if the site were above-grade, compensation would have to be made to each individual lot owner along the entire straight line from Boston to New York for the loss in their property value as a result of having the facility installed on the part that was taken.

However, if the installation was done underground, damages could be very modest indeed. While common law tells us that ownership rights exist all the way to the center of the earth and all the way to the stars, as a practical matter the approach to valuing an underground easement is to value the entire property just before the taking and to value it just after the taking in light of the fact that there is an underground easement. It is difficult to see how an easement located far underground would have any significant impact upon the value of the surface of the land if the subterranean easement is so deep that it would not interfere with the normally anticipated utility of the subject property. To the extent that hatchways and access points are required, those would require surface acquisitions at fair market value of the land acquired. To the extent that additional surface utility is required during the course of construction, such temporary easements are routinely acquired by the state and local governments to provide work stations for employees, vehicles, equipment, and supplies as the work goes forward. The essential just compensation for such temporary use is measured in terms of rental value of the land occupied for the time that it is used.

Finally, whatever opposition might exist in any of the four states involved in this project might be muted if the project were to proceed through an underwater tunnel. There is, of course, no existing straight line water route between Boston and New York although one exists between Buzzards Bay and Manhattan, and there is a straight line between Woods Hole and Miami. In light of the development that has been going on in the southerly portion of the country, perhaps a Woods Hold to Miami route would be something that the economists would find attractive.

CONCLUSION

Technologies that have evolved, are evolving, and will continue to evolve in the twenty-first century can be implemented through the use of present-day legal techniques and the application of long-established principles of law to enable them to go forward.

The creation of a private corporation armed with the power of eminent domain would enable certain acquisitions to proceed for most large-scale projects of the type envisioned in this book. Those corporations could, in fact, be made tax-exempt in all or in part and could be subject to investment and tax incentives that would attract private capital to the public service market.

Real property required for a project can be located and evaluated in a realistic and thoughtful manner, one which considers local community concerns and which takes as little as possible, for example, through the use of easements where appropriate.

Site selection for any macro-engineering project must be accomplished with a keen awareness of the environmental and economic impact of the facility on the local community in which the particular work will be done. An appropriately funded corporation endowed with the power of eminent domain can produce land necessary for any installation.

Law exists to help the national community. The law of eminent domain has historically been adapted to meet varying needs generated by modern advances.

It is available to meet those needs today.

NOTES

1. Davidson, F.P. and Meador, C.L. (eds.), *Macro-Engineering and the Future: A Management Perspective.* Boulder, CO: Westview Press, 1982.

2. Ibid., p. xix.

3. See Mansnerus, "Judicial Review in Eminent Domain", 58 *NYU Law Review* 409.

4. Davis, *Law of Compulsory Purchase and Compensation.* Surrey, England: Tolley Publishing Company, Ltd., 1994, pg. 3-8.

5. Davis, op. cit., p. 8.

Section Six

OUTLOOK

23

Macro-Projects in China

Ernst G. Frankel
Professor of Ocean Engineering, M.I.T.

INTRODUCTION

Since China's emergence from economic isolation in 1978, the economy of the People's Republic of China has grown at a phenomenal pace. It is expected that perhaps five years from now, it will overtake Japan as the world's largest economy; if relative economic growth rates between China and the U.S. remain at their current levels China should surpass the economy of America within fifteen to twenty years. This is not just due to the rapid opening up of the Chinese economy to foreign investment and international trade but also to the increasing development of China's human and natural resources.

The population of China will exceed 1.268 billion people by the end of this century -- about 24.8% of the world's population. Educational literacy stands at 97%, making it among the world's highest. The average age of the population has increased slightly to 25.4 years as a result of fairly strict population control, but the most striking demographic development is the rapid growth of China's urban centers which now account for over 26% of the population. There are 22 cities with populations of more than 4 million people, and four cities have populations in excess of 10 million.

China's natural resources include 940 billion tons of recoverable reserves of coal (with an annual production of 600 million tons) -- the world's largest proven reserves of coal. The country also has enormous hydroelectric power potential estimated at 676 million KW, of which only a small fraction has been harnessed to date. China's petroleum reserves are mainly offshore and about 100 billion barrels of recoverable oil have been proven so far.

As the world's second largest steel producer, China has to import about 78 million tons of iron ore per year to supplement 49 million tons per year produced domestically. To support China's rapid economic growth, which has averaged 11.8% per year since 1985 (about 4.2 times the world's average growth rate during the same period), China is planning a significant number of major projects (mainly infrastructure), as shown in Exhibit 23.1, which together amount to $173.8 billion in 1996 terms.

Three Gorges Dam	$78.0 B
Yangtze River - North China Aqueduct	$30.0 B
Yellow River Project	$12.0 B
Yangtze River Estuary Project	$2.0 B
Pudong Airport Project	$2.0 B
Railway Electrification	$4.0 B
Yangtze River Railway Projects	$3.0 B
Yangtze River Shipping and Port System	$4.8 B
China Port Projects: Shanghai, Ningbo, Xiamen, Yantian, Hong Kong, Dalian, etc.	$8.2 B
China Shipbuilding Projects	$4.0 B
Urban Transportation Projects	$4.9 B
Electric Power Generation & Co-generation	$11.2 B
Steel Industry	$4.1 B
Deep Water Oil/Dry Bulk Terminals	$1.0 B
Telecommunications Systems	$7.8 B

Ex. 23.1 Major Chinese Infrastructure Projects: 1996-2000

CHINA'S ECONOMIC WOES

Much has been said about China's economic problems and how they will affect the future economic growth of the country. Many projects, large and small, have failed, and many state-owned enterprises that are unable to meet the growing competition or work under a challenged environment suffer huge deficits and cannot pay their workers. Inflation, although down from well over 20% in 1994, still runs 14-15%. In general, the economy is split between a largely thriving private sector and a stagnant state or public sector. Tight credit helped reduce inflation in 1995, but choked the state sector even further.

Another issue is the increasing competition by regions, provinces, and municipalities both for state support and for foreign investment. This now pits the south against the central lower Yangtze region and both against Beijing or even the resource-rich northern part of China. Tension between the poorer inland provinces and the richer coastal provinces over growing inequalities in income and opportunity continues to haunt the Beijing leadership which wants to continue an economic path toward growth while maintaining national unity. Unity and competition among regions has become a major issue. Meanwhile foreign investment growth slowed in 1995, and declined even further in 1996, making this regional competition even keener.

It is hoped that many of the macro-engineering projects being planned will help pull the country together by providing greater opportunities over a wider area, pulling the country's economy together, and thereby building greater unity while enhancing competitiveness.

Economic development has been highly concentrated in the coastal areas. The average per capita income in the coastal areas, which comprise about 6.8% of China's land area and 24% of its population, is now over $3,600 per year or nearly four times China's average per capita income of about $1,000 per person per year.

YANGTZE RIVER RESOURCE DEVELOPMENTS

The Yangtze River basin is the area with the largest economic potential in China. Its population accounts for nearly 40% of China's total population, and its resources are enormous.

In particular, the upper Yangtze River area offers unique opportunities. Upper Yangtze River reserves in hydropower are the largest in China -- probably the world. Its potential, exploitable volume of annual generated power is 390,000 million KW and 896,600 million KWH, making up 82% and 87%, respectively, of the total river hydropower potential and 50% of the total KW and KWH in China, with the possibility of building high dam reservoirs that would yield outstanding economic benefits. According to recent surveys it is possible to set up 86 large hydropower plants with generating capacity of 2,500 million KW, including 34 plants with the capacity of one million KW, four with capacities of 5 million KW, and two with capacities of 100 million KW.

Similarly, there are enormous amounts of mineral resource reserves in the area. About 120 mineral products are known to exist and 98 have proven resources, including coal, natural gas, apatite, ilmenite, strontium, optics fluorite, vanadium, lepidolite, cadmium, mirabilite, iodine, aluminum, asbestos, platinum, mercury, beryllium, germanium, and iron, copper, manganese, lead, zinc, silver, and pyrite.

In recent years exploitation and use of some of these resources has accelerated (see Exhibit 23.2). By 1990, 42 hydropower plants of 10,000 KW or more had been built or were under construction, with a combined generating capacity of 7,690,000 KW and total annual generated power of 40,500 million KWH.

Resource	% of country total
Hydroelectric power	9%
Natural Gas	43%
Coal	10%
Pig Iron	10%
Steel	9%
Steel Products	8%
Seamless Steel Tube & Specialty Steel	25%
Aluminum, copper and Iron Alloy products	17-20%
Phosphorus	60%
Chemicals	16%
Pyrite and Soda Ash	10%

Ex. 23.2 Accelerated Production of Resources

A major problem on the upper Yangtze River is the lack of effective infrastructure. Although national infrastructure construction has improved in the past thirty years, this area falls behind in construction of this type and cannot meet the demands of economic development due to its poor foundation, backward economic development, and lack of funds and technique.

Railways. Considering the transport network in the Yangtze River basin, in 1990 there were eight main railway lines and twenty feeder railway lines (with a total of 5880 km, including electric lines of 2062 km). The density of the railway network was 55 km per 100 square km and 0.37 km per 10,000 persons, the volume of freight transport by rail was 986,800,000 tons and the turnover of freight transport was 61,400,000,000 tons km, or an average rail transport distance of 62 km.

Highways. In 1993, there were 194,500 km of highways in the area, of which 20% were super, 40% secondary, and 40% substandard highways. The density of highway networks was 18.25 km per 100 square km and 12.78 km per 10,000 persons. The volume of freight transport by road was 8,245,500,000 tons and the turnover of freight transport was 33,500 million tons km in 1990. Most vehicles had a capacity of less than 8 tons, and the average road transport distance was only 4.06 km. In other words, road transport is mainly local.

Waterways. There are 9,000 km of waterways, including 40% with a depth less than one meter, and less than 10% are navigable by 1,000+ ton ships. The main ports are Chongqing, Fuling, Wanxian, Yibin, Luzhou, and Leshan. The volume of waterborne freight transport in 1990 was 320 million tons and the turnover of freight transport is 11,290 million tons km, or freight was carried an average of 35.3 km.

Aviation. There are four major airports including Chengdu, Chongqing, Guiyang, Kunming, and more than ten feeder airports with international air routes to Hong Kong, Thailand, Nepal and more than 100 domestic routes. Two airlines (Southwest Airlines and Sichuan Airlines) serve the region. In 1990 the number of passengers transported was 2,215,000 persons, the volume of freight transported was 47,000 tons, and the turnover of freight was 61 million tons km.

Infrastructure development is the prerequisite for economic development in the upper reaches of Changjiang River which is an area with abundant resources and backward economy.

ENERGY EXPLOITATION
Speeding up the exploitation of energy is not only a major task, but the primary subject of infrastructure construction. The two work together.

HYDROPOWER
1. In the main branch of the Changjiang River from Yibin to Chongqing, it is planned to build three step-hydropower plants:

Shipong:	2.13 million KW and 12,600 million KWH
Zhuyangxi:	1.9 million KW and 11,200 million KWH
Xiaolanhai:	1 million KW and 4,000 million KWH

2. In the Jinsha River from Shigu to Yibin, the hydropower potential is about 42,300,000 KW with the possibility of setting up a high dam. A feasibility study of the Xiangjiaba and Xiluodu hubs in progress.

      Xiangjiaba: 5 million KW and 33,100 million KWH

      Xiloudu: 12 million KW and 62,700 million KWH

3. The Yalong River is rich in hydropower potential. It is planned to set up 21 step-hydropower plants, with a total generating capacity of 22,350 million KM and annual generated power of 135,800 million KWH, including:

      Ertan: 3 million KW and 18,100 million KWH (under construction and has got loan from World Bank)

      Tongziling: 400,000 KW and 2,500 million KWH (its feasibility study is in progress)

4. The hydropower potential of the Mingjiang River is 48,880,000 KW, including the key plants of Shaba and Zhipingpu. It is planned to erect 14 step-hydropower plants with total generating capacity of 3,240,000 KW, including:

      Shaba: 720,000 KW and 4,100 million KWH

      Taipingyi: 260,000 KW and 1,500 million KWH (its preliminary design has been completed)

      Zhipinpu: 600,000 KW and 3,000 million KWH (its feasibility study has been finished)

5. The hydropower potential of the Dadu River (a tributary of Mingjiang River) is 31,320,000 KW. It is planned to build 16 step-hydropower plants with total generating capacity of 17,600,000 KW and annual generated power of 100,800 million KWH, including:

      Tongjiezhi: 600,000 KW and 3,200 million KWH (under construction)

      Pubugou: 2.8 million KW and 14,100 million KWH (with better ability to regulate and protect against the flood and sand, etc.)

THERMAL POWER
Eight thermal power bases will be constructed, supported by coal exploitation.

1. The Zhijin Electric-Coal Incorporated Base supported by Zhijin Coal Miner includes 2 plants:

      Feitian Miner-Plant: 4.2 million tons of coal and 1.2 million KW of generating capacity

      Zhijin Miner-Plant: 3.9 million tons of coal and 900,000 KW of generating capacity

2. The Shuina Electric-Coal Incorporated Base support by Shuicheng and Nayong Coal Miner includes 4 plants:

      Nayong Miner-Plant: 8.1 million tons of coal and 3.6 million KW of generating capacity

Gemudi Miner-Plant: 3.9 million tons of coal and 1.2 million KW of generating capacity

Shuicheng Plant: 400,000 KW of generating capacity

Tashan Plant: 1.2 million KW of generating capacity

3. The Wianbei Thermal Plant Base, supported by the coal miners of Zhunyi, Tongzi, Jingsha, and Shanshui, includes:

Zhunyi   574,000 KW of generating capacity

Qianbei   1.5 million KW of generating capacity

4. The Qianzhong Thermal Power Plant Base, supported by the coal miners of Guiyang and Anshun, includes

Guiyang   300,000 KW of generating capacity

Anshun   1.2 million KW of generating capacity

Qingzheng   1.25 million KW of generating capacity

5. The Congqing Thermal Power Plant Base, supported by the coal miners of Natong, Shongzhao, and Shanshui, includes the plants of Louhuang, Chongqing, and Huayingshan.

6. The Chuannan Electric-Coal Inc. Base, supported by the coal miners of Junlian and Guxu, includes the plants of Huangjiaozhuang, Chuannan.

7. The Zhaotong Electric-Coal Incorporated Base supported by the coal miner of Zhaotong includes 3 plants:

Zhaotong: 30 million tons of coal and 3 million KW of generating capacity

Weixing: 1.8 million tons of coal and 600,000 KW of generating capacity

Zhengxiong: 1.8 million tons of coal and 600,000 KW of generating capacity

8. The Diandong Thermal Power Plant Base supported by the coal miner of Quijing, includes

Diandong   1.2 million KW of generating capacity

Qujing   800,000 KW of generating capacity

Yuanwei   600,000 KW of generating capacity

THE DEVELOPMENT OF TRANSPORTATION

Railway

As the key means of transportation, railroads have an important effect on the exploitation of resources in this region. With this in mind, emphasis will be placed on renovating the transportation capacity into and out of the resource-rich regions.

1. *Construction of Access Out of the Region*
   The north railway network, including the Bao-Chengdu and Xiang-Yu, bear more freight volume than any other network. New projects include the double track of Bao-Cheng, new construction on the Xian-Ankang line, and electric renovation of the section from Daxian to Congqing of Xiang-Yu, which will be built on the existing railway network.
   The south railway network includes new construction of Nan-Kun and renovation of Qian-Gui. Nan-Kun is a convenient approach to the seaport in this region, and a new access for transporting coal and phosphorus from Yunnan and Guizhou. It began construction in 1991 and was completed in 1997. The renovation of Qian-Gui was completed at the end of 1995 to increase its freight capacity to 8,000,000 tons.
   The west railway network includes new construction of Guang-Da from Guangtong station of Cheng-Kun to Xiaguan, and will extend southwest to connect with the Burma railway. In addition, it is planned to build the railway from Chengdu to Xinin and from Guangyuan to Nanzhou.

2. *Construction of Access Into the Region*
   Projects include: the electric renovation of Cheng-Kun; the new construction of Da-Cheng which has been started in 1992, the reconstruction of the section from Anbian to Shushe (364 km) of Nei-Kun, the new construction of Shuicheng-Xiayunshang, Longchang-Huangjiao (the section from Longchang to Luzhou is under construction), Emei-Yibin (178 km), and Leshan-Zigong (105 km). In addition, the construction of several local railway lines have been started or will be started soon: Puji-Leba (22.5 km), the branch line from Jingshawan to Xunsichang of Lunlian line (72.9 km), the section from Nanchuan to Shujiang of Nanchuan-Fuling (24.2 km), Gongxian-Zhoujia (23 km) and the branch line from Quijing to Enhong of En-Hong (76.5 km).

Highways
From the viewpoint of resource exploitation, highways as an auxiliary means of transportation are of major importance in the region, especially in Sichuan. A preliminary highway network has been formed, and emphasis placed on improving the grade of existing highways and developing superhighways. Based on the superhighway circle of Chengdu, Congqing, Guiyang, and Kunming, the main national superhighways will be constructed to form economic development approaches.

Water Transport
The main branch of the Changjiang River from Yichang to Chongqing will be navigable by 5,000 ton ships during the wet season after the completion of the Three Gorges Project, and the lane from Lanjiatou to Yibin will be renovated to reach the third grade.
   On the Wujiang River, the lane from Gongtan in Guizhou will reach the fifth grade and the lane from Gongtan to Fuling will reach the fourth grade with the support of several step-hydropower plants.

On the Chishui River, emphasis will be placed on the renovation of the lanes from Chizi to Chajiao in Guizhou to reach the sixth grade and the lane from Xianyutan to Hejiang in Sichuan to reach the fourth grade.

On the Mingjiang River, emphasis will be placed on renovation of the lane from Leshan to Yibin to reach the third grade.

Aviation

There will be 17 airports either constructed or renovated, including one hub airport, three artery line airports, and five main line airports.

24

Challenges for the Twenty-First Century

Ernst G. Frankel
Professor of Ocean Engineering, M.I.T.

The global village has now become the global world, a world driven by technology developed by human genius and used to its benefit while protecting the environment of the world. The major challenges for macro-engineering in the next century could be summarized as follows:

1. **Energy**
   - Space energy generation from the sun using laser or microwave transmission to earth and/or to large space stations

   - Ocean wave and current power generation

   - Use of submerged craters or fissures emitting large volumes of high temperature flows from inside the earth's crust to generate electric power

   - Efficient generation by very large hydroelectric plants on the Yangtze River, the Amazon, the Mekong, and other as yet untamed rivers

   - Clean nuclear power (fission)

   - Large wind power generators

   - Floating relocatable emergency power plants

2. **Communications**
   - High-speed digital satellite communication systems

   - Worldwide personal wireless video telephone systems

   - Large high-speed data storage retrieval and transfer systems

   - Internet libraries replacing hard copy libraries with instant access worldwide

   - Worldwide global position, routing, collision prevention, and traffic management systems

3. **Transportation**
   - Tunnel between Africa and Europe (Morocco to Spain)

   - Tunnel between new global airport at Inchon (South Korea) and Tianjin (Beijing, China)

   - Super high-speed tunnel trains

   - Supersonic intercontinental train transport

   - Space shuttle intercontinental passenger travel

   - Automated highways

   - Mid-ocean floating ports

   - Floating relocatable airports

   - A world system of storage and logistics centers

4. **Water Supply**
   - Large regional water supply systems (Yangtze River to supply arid north China); artificial manmade river in Egypt; Libya-Turkish Middle Eastern water supply network; Rhone-Algiers aqueduct

   - Regional collection and reservoir systems

   - Regional waste water collection treatment and recycling systems

5. **Health Care**
   - Large flying or floating hospital and health care centers which can be quickly repositioned to deal with epidemics and other regional care problems

   - World medication supply system

- Worldwide telemedicine based on Internet and intranet, with real video connection

- World health consortium which shares complex facilities and equipment

6. Education
- World Internet university

- Worldwide continuing education system

- Remote interactive learning to all communities to eliminate illiteracy among those born after 1998

7. Food Production and Delivery
- Use of satellite planning and modern farming to double per hectare output by 2020. Development of large food storage and logistics hubs and well-organized distribution systems.

- Development of ocean farming

8. Environmental Systems
- Environmental pollution and other impacts are seldom local and must therefore be addressed on a regional or global basis. In some areas global impact measurements and international agreements exist, but there is an urgent need for a true satellite-based international environmental management and control system.

- Some nations have developed rapid response systems to cope with environmental disasters. A worldwide complex of rapid response systems is required to assure sustenance of the globe's environment.

9. Manufacturing
- Manufacturing will become global, requiring not only global standards but global coordination, management, and logistics control. Manufacturing will be largely robotic and much may be performed in space.

The above are examples of some of the macro-engineering developments expected in the next century. These developments will have a major impact on the way and quality of life of mankind. New values will have to be established as work assumes a different role in the lives of people. Similarly, leisure will become an increasingly complex problem for people accustomed to spending most of their waking hours working, often in a physical work environment.

This brave new world may require invention of new social structures or systems, development of different interpersonal relations, and ultimately a change in setup of the world as a fractured grouping of nations, tribes, social and ethnic groups.

We will all become both more the same, yet different in a more personal sense. Good luck to mankind!

MACRO-ENGINEERING: MIT Brunel Lectures on Global Infrastructure
Editors: FRANK DAVIDSON & C. LAWRENCE MEADOR, Macro-Engineering Group, School of Engineering, Massachusetts Institute of Technology, Cambridge, Mass., USA

ISBN: 1-898563-33-X   176 pages   1997

This volume makes available the latest reflections on large-scale engineering for building a better world for tomorrow. Its interdisciplinary panel of international authorities from engineering, oceanography, academia and law describe how great and imaginative concepts may be refined, tested, adapted, financed, implemented and put to use. Here are authentic records and commentaries about some of the world's significant engineering use of software by the US military, to clean up the Gulf War oil spill pollution.

These authentic accounts of large-scale engineering achievements will convey a powerful message to world governments. It will also motivate and inspire, both academic research scientists and leaders of those industries involved, to continue striving to improve life on our planet.

Contents: Chanenel Tunnel Financial Engineering; Engineering of the Suez Canal; High Speed Transport in Europe; Prefabricated Relocatable Island Transport; Selecting Macro Projects for the Millenium; Lessons from Macro Projects; Old Cities and New Towns for Tomorrow's Infrastructure; Operation Mulberry - A Floating Transportable Harbour; Military Software in the Gulf War Oil Spill Clean-up; Fail-Safe Transport and Road Travel.

ANALYSIS OF ENGINEERING STRUCTURES
B. BEDENIK, Faculty of Civil Engineering, University of Maribor, Slovenia, *and* C. BESANT, Head of Computer-Aided Systems Engineering, Imperial College of Science, Technology and Medicine, University of London

ISBN: 1-898563-55-1   450 pages   1988

This text delivers a fundamental coverage for advanced undergraduates and post-graduates of structural engineering, and professionals working in industrial and academic research. The methods for structural analysis are explained in great detail, being based on basic static, kinematics and energy methods previously discussed in the text.

The finite element method as an extension of the displacement method is covered, but only to explain computer methods presented by use of the structural analysis package OCEAN. An innovative approach enables influence lines calculations (using ψ-*functions* developed by one of the authors) is a far simpler manner than by any previously known method. Basic algebra given in the appendices provides the necessary mathematical tools to understand the text.

Contents: Introduction; Definitions and basic concepts; Statically determinate structures; Kinematics of structures; Basic concepts of structural analysis; Deformations; Stiffness and flexibility; The force method; The displacement metho; The finite element method; Bridge analysis; Computer applications.

FINITE ELEMENT PROGRAMS IN STRUCTURAL ENGINEERING AND CONTINUUM MECHANICS

CARL T.F. ROSS, Professor of Structural Dynamics, Department of Mechanical and Manufacturing Engineering, University of Portsmouth

ISBN 1-898563-28-4   650 pages   1996

This undergraduate and postgraduate book will serve for courses in mechanical, civil, structural and aeronautical engineering; and naval architecture. It is written in a step-by-step methodological approach so that undergraduate readers can acquire knowledge, either through formal engineering courses or by self-study.

Contents: Forces in plane pin-jointed trusses; Bending moments in beams: Bending moments in rigid-jointed plane frames; Forces in pin-jointed space trusses; Static analysis of three-dimensional rigid-jointed frames; Vibration of rigid-jointed space frames; Bending moments in grillages; Vibration of grillages; Slab on elastic foundation; In-plane stresses in plates; Bending stresses in flat plates; Stresses in doubly-curved shells; In-plane vibrations of plates; Lateral vibrations of flat plates; Vibration of thin-walled, doubly-curved shells; Stresses in solids; Two-dimensional field problems; Solution of Helmholtz's equation; **Computer Programs**: Plane pin-jointed trusses; Beams "BEAMSF"; Beams on an elastic foundation "BEAMERF"; Rigid-jointed frames "FRAME2DF"; Rigid-pin-jointed plane frames "FRAMERP"; Pin-jointed space trusses; Three-dimensional rigid-jointed frames; Vibration of rigid-jointed space frames; Bending moments in grillages; Vibration of grillages; Slab on an elastic foundation; In-plane stresses in plates "STRESS4N"; In-plane stresses in plates "STRESS8N"; Stresses in double-curved shells; Vibration of in-plane plates; Lateral vibrations of flat-plates; Vibration of doubly-curved shells; Stresses in solids; Field problems "FIELD3SF", "FIELD4SF" and "FIELD8SF"; Helmholtz's equation "ACOUSTIC":

"Computer programs for finite element analysis ... students and lecturers may find them of value" - *The Structural Engineer* (Professor I.A.Macleod, Strathclyde University, Glasgow)

"These programs, written in Quick Baisc, utilize the finite element method to solve a variety of engineering problems ranging from static and dynamic analysis in two and three dimensions to two-dimensional field ... recommended to readers with a strong theoretical background, and upper-division undergraduates through to professionals" - *Choice*, American Library Association (D.A. Pape, Alfred University, USA)

ADVANCED APPLIED FINITE ELEMENT METHODS

CARL.T.F.ROSS, Professor of Structural Dynamics, Department of Mechanical & Manufacturing Engineering, University of Portsmouth

ISBN 1-898563-51-9   450 pages   1998

This text delivers a pragmatic and comprehensive coverage of applied finite elements in all the manifold aspects of engineering science, bridging the gap between traditional books in applied mechanics and specialised texts on finite element analysis. It will serve as a course text for advanced undergraduates and post graduates reading or researching in mechanical, civil, structural and aeronautical engineering or naval architecture.

Contents: Matrix algebra; Basic structural concepts and energy theorems; The discrete system; Static analysis of pin-jointed trusses; Static analysis of rigid-jointed frames; Finite element analysis; In-plane quadrilateral elements; Vibrations of structures; Grillages; Non-linear structural mechanics; Steady-state field problems; Axisymmetric problems; Transient field problems; The modal analysis method; Mathematical modelling.

CIRCUIT ANALYSIS

DR. JOHN WHITEHOUSE, Department of Engineering, University of Reading

ISBN1-898563-40-3   240 pages   1997

This text presents the fundamentals of circuit analysis in a way suitable for first and second year undergraduate courses in electronic or electrical engineering. The topology of networks is stressed, with the aid of graph theory.

Contents: Fundamentals; Network equations; Network theorems; Networks with inductors and capacitors; Network analysis using phasors; The Laplace transform in network analysis; The Fourier series and Fourier transform; The frequency response of networks; Power dissipation and energy storage in networks

MATHEMATICAL MODELLING: Teaching and Assessment in a Technology-Rich World

Prof P. GALBRAITH, University of Queensland, Australia
Prof W. BLUM, The University of Kassel, Germany
Prof G. BOOKER, Griffith University, Australia
Dr. IAN D .HUNTLEY, The University of Bristol

ISBN 1-898563-42-X   368 Pages   1997

This book contributes to the teaching, learning and assessing of mathematical modelling in an era of rapidly expanding technology. It addresses levels of education from secondary schools through teacher training colleges, colleges of technology, universities and research. The theme of technology dominates throughout in a range of applications. The use of graphics calculators, spreadsheets, symbolic manipulator software, and special purpose programs all feature.

Contents: A THEME: Issues and alternatives in assessing modelling
B THEME: Technologically enriched mathematical modelling
C THEME: Real world: models and applications
D THEME: Applications and modelling in teaching and learning
E THEME: Applications and modelling in a system or national context

TEACHING AND LEARNING MATHEMATICAL MODELLING: Innovation, Investigation and Application

Editors: PROF S.K. HOUSTON, The University of Ulster, N. Ireland
PROF W.BLUM, The University of Kassel, Germany
DR. I. HUNTLEY, The University of Bristol
DR. N.T.NEILL, The University of Ulster

ISBN- 1-898563-29-2   416 pages   1997

Contributing mathematicians from many countries across the world (Austria, Australia, Germany, Holland, Italy, Japan, Russia, Spain, UK, USA) reflect a modern and international interpretation of current knowledge and development. The interdisciplinary nature of the topics reflects application in mechanics and engineering, medicine, patient flow in hospitals, computing science, traffic control, business studies, and mathematics (fractals and analysis).

Commendation:

"Greatest value for postsecondary courses centred around mathematical applications and modelling."   *The Mathematics Teacher*
"Of great value to those teaching mathematical modelling in high schools, colleges, and universities. Recommended for libraries."
       D.V. Chopra - *Witchita State University, USA*

DIGITAL SIGNAL PROCESSING: Software Solutions and Applications

J.M. BLACKLEDGE and M.J. TURNER, Department of Mathematical Sciences, Faculty of Computing Science and Engineering, De Montfort University, Leicester

ISBN: 1-898563-48-9   *ca.* 200 pages   1998

This text is for advanced undergraduates and postgraduates reading electronic engineering, computer science and/or applied mathematics. Complete with CD-ROM it delivers the necessary mathematical and computational background and some of the processing techniques used for Digital Signal Processing (DSP). The book's appeal lies in its emphasis on software solutions for which source code is provided.

The book discusses the mathematical techniques used in signal analysis, and the theoretical methods of solution which are used to design DSP algorithms. It presents the integral transforms used for signal processing including the Wavelet transform. There is supplementary material, and a number of problems with their solutions. A range of graphical illustrations clarify the development of the subject and analysis of importance for solving the problems presented in the text. A number of worked examples are given which relate to the design and execution of DSP modules using C/C++.

Contents: PART I: MATHEMATICAL BACKGROUND - Fourier series and Fourier integrals; Convolution integrals; Analytical signals and the Hilbert transform; The sampling theorem; PART II: COMPUTATIONAL BACKGROUND - Sampling and aliasing; The convolution sum; The discrete Fourier transform; The fast Fourier transform; Computing with FFT's; Leakage and windowing; Digital filters; The FIR and IIR filter; PART III: PROCESSING TECHNIQUES - Inverse filters; The Wiener filter; Constrained deconvolution; The matched filter; Bayesian estimation; Maximum entropy filters; Non-stationary deconvolution; Super resolution techniques; Statistical filters; Singular value decomposition; The Kalman filter; Dynamic programming techniques; Fractal analysis of statistically self-affine signals; Wavelets.

DIGITAL FILTERS AND SIGNAL PROCESSING IN ELECTRONIC ENGINEERING: Theory, Applications, Architecture, Code

S.M. BOZIC and R.J. CHANCE, School of Electrical and Electronics Engineering, University of Birmingham

ISBN 1-898563-58-6   *ca.* 200 pages   1998

From industrial and teaching experience the authors provide an unusual blend of theory and practice of digital signal processing (DSP) for advanced under-graduate and postgraduate engineers readhing electronics in a fast-moving and developing area driven by the information technology revolution. This is a source book of research and development for design engineers of embedded systems in real-time computing, and applied mathematicians who apply DSP techniques in telecommunications, aerospace (control systems), satellite communications, instrumentation, and medical technology (ultrasound and magnetic resonance imaging).

The book is particularly useful at the hardware end of DSP, with its emphasis on practical DSP devices and the integration of basic processes with appropriate software.. It is unique to find in one volume the implementation of the equations as algorithms, not only in MATLAB but right up to a working DSP-based scheme. Other relevant architectural features include number representations, multiply-acccumulate, special addressing modes, zero overhead iteration schemes, and single and multiple instructions.

Contents: Basic concepts and analytical tools;   Fundamental features of DSP; Practical DSP devices;   Discrete Fourier transform; DSP implementations of Fourier and Goertzel algorithms;   FIR filter design methods;   IIR filter design methods;   Other topics in digital signal processing.

SIGNAL PROCESSING IN ELECTRONIC COMMUNICATIONS

MICHAEL J. CHAPMAN, DAVID P. GOODALL, and NIGEL C. STEELE, School of Mathematics and Information Sciences, University of Coventry

ISBN 1-898563-30-6   288 pages   1997

This text for advanced undergraduates reading electrical engineering, applied mathematics, and branches of computer science involved with signal processing (speech synthesis, computer vision and robotics). Serves also as a reference source in academia and industry.

Signal processing is an important aspect of electronic communications in its role of transmitting information, and the mathematical language of its expression is developed here in an interesting and informative way, imparting confidence to the reader.

Contents: *Si*gnal and linear system fundamentals; System responses;   Fourier methods; Analogue filters;   Discrete-time signals and systems;   Discrete-time system responses; Discrete-time Fourier analysis; The design of digital filters; Aspects of speech processing;   Appendices: The complex exponential; Linear predictive coding algorithms; Answers.

Printed and bound by CPI Group (UK) Ltd, Croydon, CR0 4YY

03/10/2024

01040434-0016